Computing in Geographic Information Systems

Computing in Geographic Information Systems

Narayan Panigrahi

CRC Press
Taylor & Francis Group
Boca Raton London New York

CRC Press is an imprint of the
Taylor & Francis Group, an **informa** business

CRC Press
Taylor & Francis Group
6000 Broken Sound Parkway NW, Suite 300
Boca Raton, FL 33487-2742

First issued in paperback 2019

© 2014 by Taylor & Francis Group, LLC
CRC Press is an imprint of Taylor & Francis Group, an Informa business

No claim to original U.S. Government works

ISBN-13: 978-1-4822-2314-9 (hbk)
ISBN-13: 978-0-367-37856-1 (pbk)

Library of Congress Cataloging-in-Publication Data

Panigrahi, Narayan.
   Computing in geographic information systems / Narayan Panigrahi.
      pages cm
   Includes bibliographical references and index.
   ISBN 978-1-4822-2314-9 (hardback)
   1. Geographic information systems--Mathematical models. I. Title.

G70.212.P35 2014
910.285--dc23   2014003932

Visit the Taylor & Francis Web site at
http://www.taylorandfrancis.com

and the CRC Press Web site at
http://www.crcpress.com

This book is dedicated to the loving memory of my parents
Shri Raghu Nath Panigrahi and Smt Yasoda Panigrahi
Village Pallipadnapur, District Ganjam, State Odisha of India
who brought me up with dedication and placed education second to none despite their modest means

Contents

List of Figures

List of Tables

Introduction

The progress of GIS (Geographic Information System) over the past two decades has been phenomenal. The quantity and quality of research literature contributed, new applications developed and systems engineered using GIS are indicators of its growing popularity among researchers, industry and the user community. Though GIS derives its acronym from Geographic Information System, it has emerged as a platform for computing spatio-temporal data obtained through a heterogeneous array of sensors from Land-Air-Sea in a continuous time frame. Therefore, GIS can easily be connoted as Spatio-Temporal Information (STI) system.

The capability of continuous acquisition of high spatial and high spectral data has resulted in the availability of a large volume of spatial data. This has led to the design, analysis, development and optimization of new algorithms for extraction of spatio-temporal patterns from the data. The trend analysis in spatial data repository has led to the development of data analytics. The progress in the design of new computing techniques to analyze, visualize, quantify and measure spatial objects using high volume spatial data has led to research in the development of robust and optimized algorithms in GIS.

The collaborative nature of GIS has borrowed modeling techniques, scientific principles and algorithms from different fields of science and technology. Principles of geodesy, geography, geomatics, geometry, cartography, statistics, remote sensing, and digital image processing (DIP) have immensely contributed to its growth. In this book I have attempted to compile the essential computing principles required for the development of GIS. The modeling, mathematical transformations, algorithms and computation techniques which form the basis of GIS are discussed. Each chapter gives the underlying computing principle in the form of CDF (Concept-Definition-Formula). The overall arrangement of the chapters follows the principle of IPO (Input-Processing-Output) of spatial data by GIS.

This book is intended to encourage the scientific thoughts of students, researchers and users by explaining the mathematical principles of GIS.

Preface

Each time I wanted to experiment and analyze the spatial data presented to me, I was confronted with many queries such as: Which GIS function will be suitable to read the spatial data format? Which set of functions will be suitable for the analysis? How to visualize and analyze the resulted outputs? Which COTS GIS has all the related functions to meaningfully read, analyze, visualize and measure the spatio-temporal event in the data?

Even if I were to select a COTS GIS system which is most suitable to answer all these queries, the cumbersome process of fetching the COTS GIS along with its high cost and strict licensing policy discourages me from procuring it. That made me a very poor user of COTS GIS and associated tools.

But the quest to analyze, visualize, estimate and measure spatial information has led me to search for the mathematical methods, formulae, algorithms that can accomplish the task. To visualize terrain as it is through modeling of spatial data has always challenged the computing skills that I acquired during my academic and professional career.

The alternatives left are to experiment with the growing list of open source GIS tools available or to design and develop a GIS software. Compelled by all these circumstances I developed a set of GIS tools for visualization and analysis ab initio.

The design and development of GIS functions need deeper understanding of the algorithms and mathematical methods inherent in the process. The first principle approach of development has its own merit and challenges. This has led me to delve into the mathematical aspects of geodesy, cartography, map projection, spatial interpolation, spatial statistics, coordinate transformation etc. This book is the outcome of the associated scientific computations along with the applications of computational geometry, differential geometry and affine geometry in GIS.

Putting all these scientific principles together I came up with a new definition. GIS is a collaborative platform for visualization and analysis of spatio-temporal data using computing methods of geodesy, photogrammetry, cartography, computer science, computational geometry, affine geometry, differential geometry, spatial statistics, spatial interpolation, remote sensing, and digital image processing.

This book is intended for students, researchers and professionals engaged in analysis, visualization and estimation of spatio-temporal data, objects and events.

Acknowledgments

First and foremost my reverence to Almighty for the blessings showered upon me. I wish to thank Professor B. Krishna Mohan, my guide and mentor for his suggestions, proofreading and encouragement.

My wife Smita is my perennial source of strength and support. She has been a constant guiding factor throughout the compilation of this book. My son Sabitra Sankalp and daughter Mahashweta motivated me throughout and made the long hours of thinking and consolidation a pleasure. Sabitra has contributed enough to understand the scientific principles of GIS and helped in proofreading some of the mathematical equations presented. The kind blessings of Shri Sashi Bhusan Tripathy and Smt Kalyani Tripathy are a boon.

Thanks to all the reviewers of this manuscript whose suggestions and new ideas have improved its quality. The suggestions of Dr G. Athithan, Outstanding Scientist, and Prof. P. Venkatachalam of IIT, Bombay are gratefully acknowledged.

This book would not have been possible without the relentless efforts of a few individuals who have contributed in many aspects to enhance the quality, including Cyan Subhra Mishra, trainee, who added all the questions and meticulously worked out the answers for each chapter and enhanced the book's relevance to the student community. He has also carefully reviewed the mathematical aspects of map projections.

The technical help rendered by my group, M. A. Rajesh, Rajesh Kumar, Shibumon, Vijayalaxmi, Jayamohan, Rakesh, Sunil and Vikash, who have gone through the chapters meticulously to avoid any typographical errors, is thankfully acknowledged.

Thanks to all my colleagues, who have encouraged me in my endeavor. I heartfully thank Mr. V. S. Mahalingam, Distinguished Scientist and ex-director of CAIR for putting the challenge before me. Thanks to Mr. M. V. Rao, Dr. Ramamurthy, Mr. C. H. Swamulu, Mr. K. R. Prasenna, Dr. Rituraj Kumar and Dr. Malay Kumar Nema.

Finally, my thanks are due to Mr. Sanjay Burman, Outstanding Scientist and Director, Center for Artificial Intelligence and Robotics (CAIR), C.V. Raman Nagar, Bangalore, for his constant encouragement and granting me permission to publish this book.

Author Bio

Dr. Narayan Panigrahi is a practicing geo-spatial scientist in the Center for Artificial Intelligence and Robotics (CAIR), one of the premier laboratories of the Defence Research and Development Organization (DRDO), Bangalore, India. He is engaged in the field of research and development of spatial information science and systems. He is the recipient of the Gold Medal for the best graduate of Berhampur University, Odisha, India in 1987. He obtained his master's degree from the J.K. Institute of Applied Physics and Technology, University of Allahabad, his M Tech from the Indian Institute of Technology (IIT), Kharagpur and his PhD from IIT, Bombay, India. His current research interests include geographical information science, design and optimization of algorithms for computation of spatial data in vector and raster form obtained through different sensors, and application of GIS for resource mapping and operation planning.

1

Introduction

Geographical Information System (GIS) is a popular information system for processing spatio-temporal data. It is used as a collaborative platform for visualization, analysis and computation involving spatio-temporal data. GIS is the name for a generic information domain that can process spatial, a-spatial or non-spatial and spatio-temporal data pertaining to the objects occur in topography, bathymetry and space. It is used for many decision support systems and analysis using multiple criteria. It has emerged as one of the important systems for collaborative operation planning and execution using multi criteria decision analysis involving land, sea and air. The popularity and usage of GIS can be judged by the large amount of literature available in the form of books [21], [31], [20], [12],[25], [4],[8], [57], [41], scientific journals such as the *International Journal of Geographical Information Science, Cartography and Geographic Information Science, Computers and Geosciences, Journal of Geographic Information and Decision Analysis, Journal of Geographical Systems, Geoinformatica, Transactions in GIS, The Cartographic Journal, The American Geographer, Auto-Carto, Cartographics* and the research publications from academic and scientific organizations. From these research literatures the growing trend in design of algorithms and novel computing technique for visualization and analysis of spatio-temporal data is evident.

GIS is evolving as a platform for scientific visualization, simulation and computations pertaining to spatio-temporal data. New techniques are being devised and proposed for modelling and computation of geo-spatial data and new computing techniques are being researched and implemented to match the increasing capability of modern day computing platforms, and ease of availability of spatio-temporal data. In computer science the word computing is an all-inclusive term for scientific methods, functions, transformations, algorithms and formal mathematical approaches and formulas which can be programmed and software codes which can be generated using high and low level programming languages. The scientific aspects of GIS are evolving as GI science [41]. Some of the computing algorithms having the capability to solve problems in different application domains are discussed in [13].

1.1 Definitions and Different Perspectives of GIS

There are different ways to describe and specify a GIS. The prime descriptive criteria of a GIS are:

1. Input domain of GIS.

2. Functional description of GIS.

3. Output range of GIS.

4. Architecture of GIS.

5. GIS as a collaborative platform for multi-sensor data fusion.

1.1.1 Input Domain of GIS

The potential of an information system in general, and GIS in particular, can be studied by understanding the input domain it can process. A review of digital data commonly available and some of the practical problems associated with directly utilizing them by GIS is discussed by Dangermond [14]. The versatility of GIS is directly proportional to the cardinality of the input domain it can process. Therefore, it is pertinent to study the input domain of GIS, i.e. the various aspects of input data such as the content of the data, organisation or format of the data, quality, sources and agencies and the way they are modeled for various uses. The input domain of an information system can be formally defined as 'the set of input data and events that it can process to give meaningful information'.

There is no empirical formula that associates the cardinality of the input domain of software to its strength and versatility; nevertheless, the anatomy of GIS can be analysed by studying the input domain of the GIS. In the next section an attempt has been made to portray the strength of GIS through its input domain. It is also important to understand the issues associated with spatial data viz. sources and agencies from where the data originates, considerations of modeling the digital data for different usage, the quality etc.

Satellite technology has brought a sweeping change to the way space imaging is done. In tandem with this progress, geo-spatial data capturing has witnessed phenomenal growth in the frequency at which the images of a particular portion of the Earth can be taken with varying resolution. In other words, the frequency (temporal resolution) of capturing spatial data has increased, and so has the spatial resolution, spectral resolution and readiometric resolution of the spatial data, i.e. the data obtained can capture in greater detail, the features of the Earth's surface. To cope with this advancement in data capturing, geo-spatial technology is trying to keep pace by providing powerful spatial processing capabilities, that can handle a large volume of spatial data for extracting meaningful information efficiently. Innovative products such as

Google Earth, Google Sky, Yahoo Street Map, WikiMapia etc. are examples of such systems that are now in common use on the internet.

A growing input domain means growing areas of applications such as location services, navigation services, area services etc. This has led to increased user domain of GIS. Therefore, GIS which earlier was largely confined to stand-alone computing platforms accessed by single user has emerged as a common resource of spatial data and computing. This has led to creation of spatial data infrastructure by large organizations. The spatial data infrastructure is accessed by large group of users through world wide web (WWW). This has led to an increased research effort for architecting and designing of efficient and multi-user GIS. The services offered by spatial data infrastructure had led to designing of enterprise GIS. Use of enterprise GIS by the internet community has pushed the research effort to integrate large volume of spatial and non-spatial data sourced from the internet users. The need for analysis of the crowd sourced data in the spatial context has pushed the geo-spatial community to evolve an innovative set of techniques known as the spatial data fusion and spatial data mining techniques. Choosing the appropriate geo-spatial data from a spatial database for a specific application then becomes an issue. The issues that need to be resolved are:

1. What spatial data formats to choose.

2. What is the geodetic datum to be used?

3. What should be the coordinate system of the data?

4. What map projection is suitable for the data?

5. Which geo-referencing method or map projection method is to be applied on the data?

The answer to all such queries can be resolved by careful study of the input domain and associated metadata. This calls for creating a database of metadata of the available geo-spatial data. Designing of a database of metadata of the spatial data resources has become a national concern. This is discussed in detail at the end of this chapter. To analyze various aspects of the input domain, it has been listed in a tabular form along with their content and format.

The broad specification of the inputs processed by a GIS along with their formats and the topology are listed in Table 1.1. This is an example set of inputs to GIS and by no means exhaustive and complete. The input domain of GIS is ever increasing and augmented because of emerging new GIS applications and creative products.

1.1.2   Functional Profiling of GIS

GIS can be considered as a set of functions which are program manifestation of algorithms in a computing platform. The set of algorithms or functions act on the spatial data (input domain) and transform them through computations

Input Data Type	Source	Topology / Format
Raster scanned data	Scanner, unmanned aerial vehicle (UAV), oblique photography	Matrix of pixels with the header containing the boundary information, GeoTIFF, GIF, PCX, XWD, CIB, NITF, CADRG
Satellite image	Satellite	BIL, BIP,BSQ
Vector map	Field survey, output of Raster to Vector (R2V) conversion through digitization	DGN, DVD, DXF, DWG
Attribute data	Field survey, statistical observation, census data	Textual records binding several attribute fields stored in various RDBMS e.g. Oracle, Sybase, PostgreSQL etc.
Elevation data	Sensors, GPS, DGPS, LIDAR, RADAR, hyperspectral scanner, digital compass	Matrix of height values approximating the height of a particular grid of Earth's surface. DTED-0/1/2, DEM, NMEA, GRD, TIN
Marine navigation charts or bathymetric charts	Marine survey, coast and island survey, hydrographic and maritime survey through SONAR	S52, S57, S56, S63 electronic navigation charts, coast and island map data
Ellipsoid parameters, geodetic datum, geo-referenced information, coordinate system information	Geodetic survey, marine survey, satellite based measurements through laser beams, geodatic triangulations	Topology: semi-major axis, semi-minor axis, flattening/eccentricity, origin of the coordinate center, the orientation of the axis with respect to the axis of rotation of Earth, Earth centered Earth fixed reference frame
Projection parameters	Geodetic survey organisations or agencies	As metadata or supporting data to the main spatial data-often saved as header information of the main file
Almanac and metrological data	Almanac tables	Time of sunrise, sunset, moon rise, moon set, weather information including day and night temperature and wind speed etc.

TABLE 1.1
Input Domain of a GIS

to various outputs (Output Range) required by the user. To facilitate this process of transforming the inputs to outputs, the user interacts with the GIS system through a GUI (Graphical User Interface) selecting different spatial and non-spatial data, value of parameters and options.

To understand the GIS functions, profiling its macro and micro functions gives an indepth processing capability of the GIS. The exhaustive set of functions in a GIS gives the cardinality of its computing capability. Further efficacy and strength of each of its functions can be measured by analyzing the order of space and time complexity of corresponding algorithms. In a way GIS can be defined through the following empirical equations:

Output Range $\leftarrow$ GIS(Input Domain)
GIS $= \{F_i : i=1,2,...,n$ is a set of n functions$\}$
$F_i = \{$Stack of Algorithms$\}$
$Y_i \leftarrow F(X_i)$

where X_i can be an atomic spatial data or set of spatial data in the form of pixels in case the input is an image or vector elements describing spatial objects such as points, lines and polygons etc.
In the above equation, function $F(X_i)$ is an algorithm if F has the following properties:

- Finiteness; i.e. it must act on the data through a finite set of instructions and complete computing in a finite time.

- Definiteness; i.e. it must result with a definite output.

- Input; i.e. the function must take some tangible input data.

- Output; i.e. the function must generate tangible output as result.

- Processing; i.e. the function must transform the input data to output data.

Therefore if the F satisfies the above conditions then it can be considered as an algorithm A and the equation can be rewritten as:

$$Y_i \leftarrow A(X_i) \tag{1.1}$$

In a sense the set of algorithms in a GIS can be thought of as the kernel of the GIS. They can manifest in the form of software components such as classes, objects, Component Object Models (COM), and Distributed Component Object Models (DCOM). The interfaces of these software components are sets of API (Application Program Interfaces) which are exposed to users or programmers to customize the GIS according to the requirements of various systems. A macro functional view of a GIS is depicted in Figure 1.1. Often the GIS algorithms act on the spatial data sets sequentially or in a cascaded manner or concurrently. Sometimes the output of one algorithm can be input to the next algorithm. This can be depicted in the following meta equation.

$$Output \leftarrow A_n O A_{n-1}......A_1(Input) \tag{1.2}$$

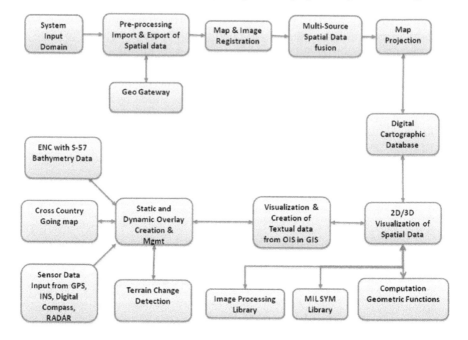

FIGURE 1.1
Block diagram depicting the macro GIS functions

As an example, computing the shortest path between start location and destination location, the computing steps can be performed through the following series of algorithms:

Input = {Source Location, Destination Location, Vector Topo Sheet with Communication Layers, Digital Elevation Model Corresponding to the Topo Sheet, Weather Information, Attribute Data}

- **Step 1**: Read the vector data and extract the communication Layer from it.
- **Step 2**: Generate a DAG (Directed Acyclic Graph) from the communication layer.
- **Step 3**: Store the DAG in a 2D array or a linked list.
- **Step 4**: Apply Dijkstra's shortest path algorithm.
- **Step 5**: Display the shortest path and all paths in user defined colours on the map and store the shortest path as a table.

One can observe from the above sequence of steps that the first two steps are pre-processing of the spatial data and are input to the main computing algorithm 'Dijkstra's Shortest Path' algorithm. The outputs are both visual and numerical. There can be many choices to the main algorithm in the form of A-star algorithm, Belman Ford, Ant Colony Optimization etc.

Therefore, to understand and specify the computing capability of a GIS it is important to profile its functional capability and the crucial algorithms used to realize them. A naive functional description or macro functional description of GIS is given in Figure 1.1. Each of these functional blocks can be further analyzed to trace the atomic or micro functions and algorithms.

Each chapter in this book has taken up the macro functions and the computing principle behind the function is described. Although this cannot claim to be complete, the most frequently computing method is discussed in each chapter.

1.1.3   Output Profiling of GIS

Output profiling of GIS is important to understand the cardinality of its computation power. The application of GIS depends on its output range. Because of rapid research and development in spatio-temporal processing methods, the output range of GIS is ever increasing and so is its application domain. Therefore it is naive to profile all the output that the GIS system can produce. Nevertheless the GIS outputs can be listed by categorizing them into the following:

1. Preliminary outputs of GIS are the visualization and measurements of spatio-temporal objects produced by GIS.

2. Secondary outputs of GIS can be computed or inferred using the preliminary outputs. These are the analytical outputs of GIS.

3. The visual output, visual and numeric simulations performed by GIS can be termed as the extended output of GIS.

1.1.4   Information Architecture of GIS

Architecture is an important design artefact of any system. Architecture gives the skeletal picture of the assembly and subassembly of the entire system. Therefore to understand the design of any information system in general and GIS in particular it is important to understand its information architecture. Information architecture gives the flow of information in the system. The flow of information from the database to the user through a series of request- response cycles. The traversing of the information from the database through software, network, hardware and finally to the GUI of the user is decided by the architecture of the GIS system. Architecture plays a crucial role in the way the user utilizes the service of a system. The information architecture decides the response time of the system, the reliability of its services, its efficiency etc. Architecture is always described through examples such as architecture of the city, temple, building, land scape, information system etc. Therefore architecture is a design artefact which is a pre or post qualifier of any major system. It is important to put a quasi definition of architecture for completeness.

Architecture is an artefact or formal specification of the system and subsystems describing the components, their topology and interrelationship in the overall system.

In software engineering paradigm generally architecture is designed, components are implemented, assembled or made and the system is integrated. Therefore design is considered as a highly skilled engineering task compared to development and integration.

GIS is a major driving force for innovation in information architecture. Architecture plays a crucial role in deciding the end user application or system where GIS is a component or system in itself. Architecture plays a crucial role to convey and express the idea of constructing the components, subsystems and the overall system to the actual builders and developers of the system. Also it is a guiding factor to make a set of developers to develop a system coherently so that the individual components can be integrated smoothly to get the end objective. It helps to convey the overall idea of the system to the developers, financers, engineers and integrators. Also the architecture is important to market the system in the post development scenario.

1.1.4.1   Different Architectural Views of GIS

GIS was in its infant stage when large computing machines such as main frame systems were the main source of collaborative computing. The emergence of desktop architecture has enabled a platform for desktop GIS, where a privileged user uses the complete computing resources to analyze and visualize spatial data stored in the local hard disk drive. This system suffers from limited usages and exploitation of GIS. Sometimes the services are denied because of system down- time due to wear and tear or a complete crash of the system. Often it is very difficult even to retrieve the high valued and strategically important data. Therefore to minimize the above limitations and to maximize the utilization of the services of the overall system client server architecture in hardware emerged. GIS graduated to adopt to the client server architecture allowing multiple users to access a centralized database server holding spatial data. In this scenario both the data and GIS server are collocated in a central server where a set of common user access the services through a network environment. Client server architecture has advantage of maximizing utilization of the GIS computing and spatial data resources by a set of close user groups through a Local Area Network (LAN). However the client server GIS suffers from the following anomalies:

1. The request-response cycle of the user for fetching services gets unexpectedly delayed when the system is accessed by the maximum

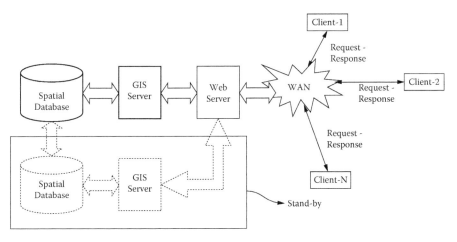

FIGURE 1.2
Multi-tier architecture in GIS

number of users. Therefore the load balancing of the database server as well as the GIS application server gets degraded.

2. A malicious user can cripple the database as well as the GIS server through access points leading to unnecessary denial of services to the genuine users.

3. Poor utilization of computing as well as the spatial data services.

Therefore to overcome the above anomalies, multi-tier architecture GIS has evolved. Multi-tier information architecture is an information architecture deployed in a LAN or a Wide Area Network (WAN) where the backend database server is abstracted from the users. The three-tier architecture is particular instance of a multi-tier information architecture. The typical configuration of a three-tier architecture GIS is depicted in Figure 1.2.

One can observe from the block diagram that the database server which is a backend server holding the crucial spatial data is abstracted from the users by a web-server and a GIS application server. The direct request for computing, visualization or analysis service by users are handled through a series of request-response cycles.

The typical sequence a request-response cycle is:

1. User's request for services is channelled through the web server.

2. Web server in turn requests the spatial computing service from the GIS application server.

3. The GIS application server requests the spatial data required from the backend data server.

4. The back end database server responds with appropriate spatial data back to the GIS application server which processes and computes as per the request of the user.

5. The processed spatial information is sent as a response through the web server to the appropriate request through the browser.

Therefore in the multi-tier information architecture, the GIS is referred as enterprise GIS or Web GIS or three-tier GIS depending on its configuration and usage. This has overcome the following limitations of the client-server architecture:

1. Abstraction of the spatial data server from the direct intervention of the user protecting it from malicious attacks.

2. The GIS application tier and the database tier can be mirrored or the multiple instances of the systems can be configured in the network to act as hot-standby or as disaster recovery facility which can address the load balancing of user requests for performance optimisations and for enhancing the reliability and availability of the services to enforce measured usages by authorized users of data and services.

3. Multi-tier GIS helps to leverage the GIS services by a vast community of users through a WAN geographically spanned across the globe. Some of the active examples of enterprise GIS are Google Earth, Google Sky, Yahoo Street Map, WikiMapia.

In multi-tier architecture GIS the clients can be classified as:

1. Rich or thick clients which enjoy high bandwidth services between the clients and server. Rich clients are privileged to access GIS service requiring high data transfer rate, high degree of computing function such as 3D terrain visualization, fly through and walk through simulations etc.

2. Thin clients are those who use generic GIS services such as map visualization, thematic map composition and measurement services.

The three-tier GIS architecture is a popular architecture which has overcome most of the limitations of the client server architectures.

With the high availability of processed spatial data and low cost spatial computing devices and sensors the services of GIS are becoming rapidly popular by large sets of users empowered with low cost computing and communication devices such as cellular phones, tablet PCs etc. The popular GIS services available through mobile devices include location based services, navigation services, measurement services, query services, weather information services, traffic information services, facility location services etc. This has led to the emergence of service oriented architecture where an atomic service can be defined as a self contained process embedding the request and response cycle with the spatial data and requested information in a single control thread. GIS enabled by the service oriented architecture has created a vast global user community spanning across land, air and sea. Therefore from the inception of GIS to its current state it has enabled and driven research in the evolution of information architecture. In some cases GIS has meta-morphed itself into systems amenable to the architecture. In some cases it has put the challenge to the research community for evolving architectures and computing paradigms. The computing paradigms such as distributed computing, grid computing, cloud computing, network computing, and quantum computing make their impact in GIS by evolving new algorithms, processing large volumes of spatial data.

1.1.5 GIS as a Platform for Multi-Sensor Data Fusion

GIS inherently collates, collects, processes and disseminates processed spatio-temporal data and information. Sensors such as Global Positioning System (GPS), Differential Global Positioning System (DGPS), Light Detection and Ranging (LiDAR), Radio Detection and Ranging (RADAR), Sound Navigation and Ranging (SONAR), digital compass, multi and hyper spectral scanner etc. basically produce data about the location, speed, direction, heights etc. of the spatial objects. Therefore the common basis of outputs of all these sensors are spatial coordinate and geometric measurements. Often these sensors are placed and operated through some platforms such as satellites, Unmanned Aerial Vehicle (UAV), Unmanned Ground Vehicle (UGV), ships etc. Therefore GIS becomes a platform for bringing these data to a common frame of reference for understanding, visualization and analysis. The common frame of reference can be imparted using GIS by modeling these data to a common datum, coordinate reference system and cartographic projection before displaying them in a digital container such as Large Screen Projection (LSP) system or computer screen. This process of bringing them into a common frame of reference is often referred to as Multi-Sensor Data Fusion (MSDF). Algorithms are embedded in GIS to read these sensor data online and process them for producing a common sensor picture.

The next level of MSDF is replacing the less accurate, less resolved attribute of the spatial object captured by one sensor by more accurate and high resolution attributes captured by other sensors, thus improving the resolution and accuracy of the data. The processing of the spatial data to improve its resolution and accuracy using the other sensors is the next level of sensor data fusion. In fact multiple levels of sensor data fusion technique to collate and improve the visualization of spatial data exist in the sub domain of MSDF.

Because of the spatial data handling capacity of GIS, it has emerged as

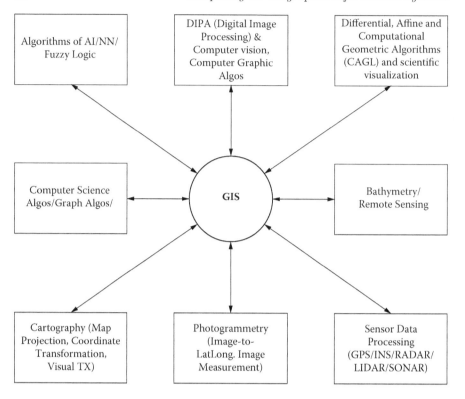

FIGURE 1.3

Collaborative diagram depicting various contributing branches of science and technology; GIS as a platform for scientific computing

the natural choice as a platform for MSDF providing a Common Operating Picture (COP) for sensors.

1.1.6   GIS as a Platform for Scientific Visualization

GIS has emerged as a platform for simulation, modelling and visualization of spatio-temporal data (Figure 1.3). It is used for visualization of scientific and natural phenomena occuring in the spatial extend. Therefore GIS can be thought of as a scientific computing platform for visualization and simulation of natural, manmade and scientific phenomena. In achieving the scientific modelling and computations, the algorithms and computing techniques from different streams of studies have contributed to GIS. Algorithms from the fields of computational geometry, differential geometry, affine geometry, remote sensing techniques [19], bathymetric, cartographic techniques such as map projection, coordinate transformations, geodesic computations, and photogrammetric technique forms the core set of computation sin GIS. Algorithms form DIP (Digital Image Processing) [26], Artificial Intelligence (AI), Neural Network (NN), fuzzy logic, computer science, graph algorithms, dynamic programming etc. help in inferential computations in GIS.

1.2   Computational Aspects of GIS

Geometric quantities pertaining to Earth such as geodesic distance, eccentricity, radius of Earth etc. form the study of geodesy. Measurement of these parameters requires a formidable mathematical modelling. The mathematical modelling of the shape of Earth gives the datum parameters. The datum parameter is associated with a reference frame so as to compute the geometric quantities associated with Earth's objects. Imparting a frame of reference to Earth's surface for deriving topological relationship among the spatial objects on Earth surface is facilitated by coordinate systems such as ECEF (Earth Centered Earth Fixed), EC (Earth Centered), ITRF (Inertial Terrestrial Reference Frame) etc.

The coordinate system, coordinate transformation, datum transformation etc. are basically mathematical formulae which play a crucial role in referencing each and every location of Earth. These transformations when repeatedly performed for a large quantity of contiguous spatial data are implemented through computing functions.

Cartography, or the art of map making, is crucial to transform the 3D spherical model of Earth to a 2D flat map. This is achieved through map projection. Map projections are mathematical transformations which transform the 3D Earth model to a 2D Earth model keeping the topological and geodesic relations among spatial objects intact. The outputs of map projection in digital form are the input to a GIS. Hence datum, coordinate system, geodesy and map projection acts as the pre-processing computations of GIS [45]. Computer cartography or digital cartograph is instrumental in the rapid progress of Geographical Information Systems [11]. Different map projection techniques their mathematical derivations and applications are discussed thoroughly by Snyder in [52], [54], [53], and a compendium of map projections have been discussed in the book entitled cartographic science by Dr. Donald Fenna [15].

Therefore datum, coordinate system, geodesy and map projection form the formidable mathematical basis providing the fundamental geo-spatial data which are the basic inputs to a GIS.

The processing of spatial information essentially can be categorized into three geral categories viz. visualization, measurement and analysis of spatial data. Visualization can be in the form of two dimension (2D), two and half dimension ($2\frac{1}{2}$) and three dimension (3D), Digital Elevation Mode l(DEM), Triangular Irregular Network (TIN), Sun Shaded Relief (SSR) model, or colour coded chloropath etc. Terrain modeling using dominant points extracted from DEM has been discussed by Panigrahi et. al. in [42].

Measurement of spatio-temporal objects or phenomena results in numeric quantities associated with spatial objects. The spatial measurements are computation of location in the form of geographic coordinate (latitude, longitude) or rectangular coordinate (Easting, Northing). Computing the distance between two locations has different connotations such as geodesic distance, planar distance, cumulative distance of intermediate measurements of consecutive locations or shortest distance etc. The direction measurements can be thought of as direction measurement from true north or with respect to magnetic north.

GIS involves computations of derived geometric quantities such as slope, aspect, area, perimeter, volume and height of spatial objects. Such quantities sometimes characterize a general area or surface giving a meaningful description of the spatio-temporal object under consideration. These quantities also have different connotation under different datum and coordinate systems. Therefore the mathematical computation of these parameters forms a crucial computing element in GIS.

Often these measurements are well formed mathematical formula applied to spatial data. Not very often the spatial data describing the spatial objects are simple rather the data may be complex and degenerate. They may not fit to a set geometric pattern. Therefore measuring these quantities requires a step by step process resulting in an algorithm. The algorithms for processing spatial data are designed keeping in mind the geometry of the data and the structure of the data. Therefore the algorithms of computational geometry, differential geometry, affine geometry and projective geometry play a critical role in processing and analyzing the spatial data.

1.3   Computing Algorithms in GIS

Table 1.2 gives a glimpse of algorithms, modelling techniques and transformations used in GIS and their possible applications in GIS. These are merely a candidate set of computing techniques commonly found in many GIS and used for computation of spatial data. Often there are multiple techniques to compute the same spatial quantity and they are dependent on the type and format of spatial data used for computation.

1.4   Purpose of the Book

GIS can be perceived as a bundle of computing techniques, modelling, and transferring, projecting, visualizing and analyzing spatial data successively.

Name of Computing Algorithm	Usage in GIS
Computing of eccentricity 'e', flattening 'f' given the semi major and semi minor axis of an ellipse	Geodesic modelling of Earth
Computing radius of curvature at prime meridian and prime vertical of Earth	Geodesy
Computing (latitude, Longitude) at any point on the surface of Earth	Geographic mapping and navigation
Computation or (Easting, Northing) given (latitude, Longitude) on a map	Map conversion, Navigation
Transformation of coordinate from spherical2rectangular, spherical2cylindrical, spherical2polar, polar2rectangular WGS84-to-NAD27, WGS-2-UAD, Molodonsky's datum transformation etc.	Coordinate transformation from one system to another, datum transformation
Map projection such as UTM, LCC, spherical, cylindrical, planar, orthographic, perspective, genomic etc.	Projection of cartographic data for preparation of maps and digital display
Computation of height, distance, direction, perimeter, area, volume, slope, aspect, curvature	Measurement services
Computation of shortest path, optimum path, critical path, alternate path from source to destination.	Operation planning and navigation
Creation of thematic maps, colour coded elevation, shaded relief map etc.	Map analysis and thematic map composition and analysis
Visualization of 3D perspective view, orthographic view etc.	Terrain visualization and analysis
Fly thru, walk thru, see thru of terrain surface	Simulation and modelling of terrain visualization and analysis
Computation of deterministic statistical methods such as mean, mode, median, standard deviation, kurtosis and inferential statistical methods such as Pearson's correlation coefficient 'r', Moran's 'I', Gerey's 'G' etc.	For statistical analysis of spatial data and interpolation of trends from the surveyed population.
Spatial interpolation methods such as Inverse Distance Weight method (IDW), Triangular Irregular Network (TIN), Voronoi's Polygon, kriging etc.	For interpolation of spatial data from the surveyed population.
Computation of line-line intersection, Point inside triangle, point inside polygon, point inside sphere, convex hull computation, Delaunay triangulation, Dirichlet's Tessellation etc.	Spatial and geometric queries. Computation of geometric quantity from spatial data

TABLE 1.2

Computing Algorithms and Their Usage in GIS

The core computing modules that make GIS are geodesic modelling which models planet Earth, sea and space to an approximate geometry. The outcome of this modelling is a mathematically defined surface or modelled surface. All the spatial objects are located, measured and referenced with respect to the modelled surface. The next set of computing involved in GIS is imparting a suitable 'frame of reference' or 'coordinate system' to the datum surface. There are different types of coordinate systems designed and being used in different datum surfaces and different applications. In GIS often there is a need to transform spatial data captured using one datum and coordinate system to another datum or coordinate system for the sake of convenient visualization, analysis and measurement. Therefore 'coordinate transformation' is a set of mathematical transformations performing the task.

Cartographic map projection is a set of geometrical transformations which intake the spatial data, datum and coordinate parameters and project the spatial data to a regular geometric surface for preparation of a map.

Computing physical parameters of spatial objects is important to give a descriptive measure of the spatial object. A set of mathematical formulae which compute the location, height, distance, direction, perimeter, area, volume, slope, aspect, curvature etc. of a physical surface or object are important constituents of GIS measurement. These descriptive quantities form the primary quantitative output of any GIS.

A distinct set of computations and mathematical formulae from differential geometry, affine geometry and computational or combinatorial geometry are used for image registration [22], geometric computations [42], [28] using structured spatial data in the form of DTED (Digital Terrain Elevation Model) or raster satellite image. The predictive logic of spatial and geometric queries is being performed on spatial data using computational geometric methods.

Spatial statistics are an emerging set of computational statistics consisting of both predictive and inferential statistical functions. These statistical functions can infer subtle trend about spatial data population from the observed samples. They can act on bivariate and multi-variate data for geospatial analysis. Spatial interpolation techniques are a set of mathematical techniques for interpolation and extrapolation of spatial data from sparsely observed data sets. IDW (Inverse Distance Weight), kriging, triangulation, tessellation etc. are popular spatial interpolation techniques with variants for different varieties of spatial data. Beside the above computations there is a set of hybrid computing techniques which are popularly used in GIS for application specific analysis and visualization of spatial data.

The application specific computations in GIS, the spatial interpolation techniques, spatial statistical methods and spatial analysis methods are emerging and therefore I felt it apt to compile these techniques and name them under 'Computing in Geographic Information Systems' or more appropriately 'Computational GIS'.

1.5   Organization of the Book

This book deals with the computational aspects of GIS. Chapter 2 delves into the mathematical modelling of Earth through geodetic datum and geodesic measurements. Chapter 3 defines the characteristics of a reference system and the usefulness of a frame of reference or coordinate system is discussed. Chapter 4 discusses some of the established frames of references being used in GIS, geodesy and spatial information systems. Chapter 5 is about cartographic modelling and transformation projections. Chapter 6 deals with the mathematical aspects of cartographic transformations in the form of different map projections. In Chapter 7, the mathematical formulae used for measurement of spatial objects and useful computing techniques used in GIS are discussed.

Chapter 8 discusses the useful differential geometric methods for computation of spatial attributes. Chapter 9 discusses the Computational geometric algorithms and their applications in GIS. The affine property, transformations and their applications for image registrations are discussed in Chapter 10. Spatial interpolation techniques are a set of tools useful for computing and generating missing spatial data from pre-surveyed data. Spatial interpolation techniques are discussed in Chapter 11. Chapter 12 discusses some of the spatial analysis techniques useful for multi-criteria-decision-analysis and used in spatio-temporal decision support systems.

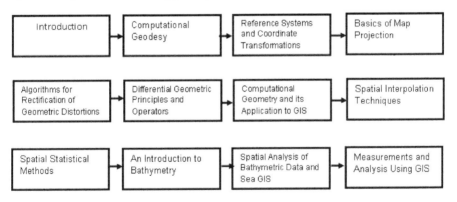

FIGURE 1.4
Organization of chapters

The book consists of 12 chapters. The organization of the chapters is given in Figure 1.4. Chapters 1–4 deal with the modelling and pre processing of spatial data and prepare the spatial data as input to the GIS system. Chapters 5–8 describe the various techniques of computing the spatial data using different geometric and statistical techniques. Finally Chapters 9–11 describe the technique for image registration computation and measurements of spatial objects and phenomena.

Detection and analysis of change in terrain using multi-dated satellite images of the same area is an important and popular GIS function. This is also a candidate application where almost all the computing techniques used in GIS are applicable. Therefore to give full details of these techniques the change detection of terrain is described in the last chapter as an application of these computational methods.

1.6   Summary

This chapter introduces GIS and its computing aspects. To encourage the interest of the reader, the information is explained through a pattern of CDF (Concept-Definition-Formulation). Multiple definitions of GIS from different perspectives are discussed. The IPO (Input-Processing-Output) pattern of the GIS as a system is analyzed. The analysis of its input domain, processing capabilities and output range are carried out through examples. The system concept of GIS as a platform for multi-sensor data fusion or platform for integration of sensor data has been explained through an example. The scientific visualization capability of GIS brings different fields of science and technology together in collaboration to make use of the spatio-temporal processing capabilities of GIS. This concept has been explained through a block diagram. Also the spatio-temporal integration capability of GIS makes it a platform for collaborative processing. The computational aspects of GIS have been explained through some candidate algorithms often used in GIS. Finally, the organization of the book has been depicted through a block diagram which essentially describes a MVC (Mode-View-Compute) philosophy of this GIS book.

2

Computational Geodesy

Mathematical modelling of the shape of the Earth and computational methods to measure its shape parameters is known as geodesy. In other words geodesy models the overall shape of the Earth through mathematical modelling. The physical properties exhibited by Earth such as its gravity field, magnetic field, polar and Keplarian motion etc. forms the basis for modelling the shape of Earth. In recent years direct measurement of the diameter of the Earth by focusing the laser beams from the satellite gives a credible accuracy of the estimations arrived by geodesic estimations. A glimpse of various applications of geodesy and its principles in measuring the physical parameters of Earth is given to make the reading of the subject interesting. A formal definition of geodesy can be the mathematical modelling of the shape of Earth and measurement of its geometrical parameters by modelling the physical phenomena exhibited by Earth such as its magnetic and gravity fields. The direct measurement of the shape of Earth is performed by focusing of laser beams fitted on the satellites revolving around the Earth. In other words geodesy is modelling, measurement and study of the physical shape of the Earth, its geometry through laser range finder, magnetic and gravity field modelling.

This chapter starts with the definition of geodesy followed with the concepts of physical, geometric and satellite geodesy. The concepts of geoid and ellipsoid are discussed. Important physical parameters of Earth such as semi major axis (a), semi minor axis (b) flattening (f) and eccentricity (e) are defined. Measurements such as radius, radius of curvature, perimeter, area and volume with respect to defined shape parameters of Earth are illustrated.

2.1 Definition of Geodesy

The dictionary meaning of geodesy is 'Dividing the Earth and measurement of the Earth'. For our purposes, geodesy is the science dealing with the techniques and methods for precise measurements of the geometry of Earth surface and its objects. The theory of modelling the shape and size of Earth through different mathematical methods such as determination of the radius of the Earth at any location, curvature measurement and computation of geodatic datum forms an important part of geodesy. Methods for measuring the spatial locations of the objects on the surface of Earth form a part of geodesy. Geodesy can be broadly classified into three branches:

1. Physical geodesy

2. Geometric geodesy

3. Satellite geodesy

Physical geodesy deals with computational techniques which allow the solutions of geodetic problems by use of precise measurement of Earth's shape through astro-geodetic methods. This method uses astro-determination of latitude, longitude, azimuth through geodetic operation such as triangulation, trilateration, base measurement etc. These methods may be considered as belonging to physical geodesy fully as much as the gravimetric methods.

Geometry geodesy appears to be purely geometrical science as it deals with the geometry i.e. the shape and size of the Earth. Determination of geographical positions on the surface of Earth can be made by observing celestial bodies and thus comes under geodetic astronomy, but can also be included under geometric geodesy.

Earth gravity field is an entity and is involved in most of the geodetic measurements, even the purely geometric ones. The measurements of geodetic astronomy, triangulation and levelling, all make essential use of plumb line being the direction of gravity vector.

Thus, as a general distinction astro-geodetic methods come under geometric geodesy which uses the direction of gravity vector, employing geometrical techniques, where as the gravimetric methods come under physical geodesy, which operates with the magnitude of 'g' using potential theory. A sharp demarcation is impossible and there are frequent overlaps.

Satellite geodesy comprises the observation and computational techniques which allow the solution of geodetic problems by the use of precise measurements to, from or between spatial locations, mostly near the Earth's satellite.

2.2   Mathematical Models of Earth

To understand how the size and shape of the Earth varies, three reference surfaces are studied widely by geodesists. The models of interest to geodesists are:

- Physical surface of Earth
- The reference geoid
- The reference ellipsoid

Mathematically describing the reference surfaces such as geoid and ellipsoid are fundamental to the understanding of geodesy.

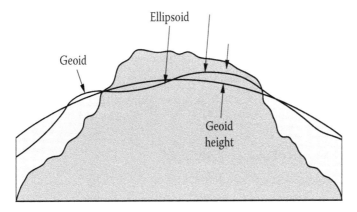

FIGURE 2.1
Separation of geoid and ellipsoid undulation

2.2.1 Physical Surface of Earth

The physical surface of the Earth is with all undulations (mountains and depressions). This is roughly an oblate ellipsoid (obtained by revolving an ellipse about its minor axis). It the actual topographical surfaces on which Earth measurements are to be made. It is not suitable for exact mathematical computation, because the formula which would be required to take the irregularities into account would necessitate a prohibitive amount of computations.

2.2.2 The Reference Geoid

On the other hand, the ellipsoid is much less suitable as a reference surface for vertical coordinates (heights). Instead, the geoid is used. It is defined as that level surface of the gravity field which best fits the mean sea level, and may extend inside the solid body of the Earth. The vertical separation between the geoid and particular reference ellipsoid is called geoidal undulation and is denoted by n. Figure 2.1 represents the ellipsoid, geoid and geoidal undulation.

The geoid is an equipotential surface of Earth's attraction and rotation. It is nearly ellipsoidal but a complex surface. The geoid is essentially mean sea level, i.e., it may be described as a surface coinciding with mean sea level in oceans and lying under the land at the sea level to which the sea would reach if admitted by small frictionless channels. The geoid is a physical reality. At sea level (geoid) the direction of gravity and axis of a level theodolite is perpendicular to it.

The mean sea level (MSL) or geoid is the datum for height measurement. The geoid may depart from ellipsoidal shape by varying amounts, up to 100m or even more.

The MSL differs from the geoid due to the following reasons:

1. The MSL surface is overlaid by air, whose pressure varies. It is not quite a free surface.

2. The wind applies horizontal force to the surface.

3. The density of water varies with temperature and salinity.

4. The sources of water, rain, river and melting of ice, do not coincide with the areas where water is lost by evaporation.

2.2.3   The Reference Ellipsoid

The ellipsoidal surface is smooth and so is convenient for mathematical operations. This is why the ellipsoid is widely used as the reference surface for horizontal coordinates in a geodetic network. It is a mathematical surface with an arbitrarily defined geometrical figure. It is closely approximate to the geoid or actual topographical surface. Since the reference ellipsoids are smooth mathematical surfaces (user defined), computations are quite easy to be performed on this surface. It is an ideal surface for referencing the horizontal position of the points on the surface of the Earth. The reference ellipsoids are of two categories.

1. In the first category the reference ellipsoid is chosen in such a way that the center of gravity of the actual Earth coincides with the center of the ellipsoid. This type of ellipsoid is called a geocentric ellipsoids, for example WGS-72, WGS-84 etc.

2. In the second category, the ellipsoids are chosen in such a way that it fits with the local datum of interest as closely as possible. For example Everest, Bessel etc.

Though several reference ellipsoids are used in the world (see Table 13.1) in the Indian context, most of the computations are performed using Everest ellipsoid on which the topographical map coordinates are referenced. For satellite based measurements and for computing the location of spatial objects on Earth's surface for scientific computations WGS-84 (World Geodatic Survey 1984), which is a geocentric ellipsoid, is used.

2.3   Geometry of Ellipse and Ellipsoid

It is important to revisit some of the geometric concepts and algebraic relationships of ellipse. The equation of ellipse with semi major axis 'a' and semi minor axis 'b' is given by

$$\frac{x^2}{a^2} + \frac{y^2}{b^2} = 1 \tag{2.1}$$

where, $b^2 = a^2(1 - e^2)$ and 'e' is the eccentricity, $0 < e < 1$. In ellipse the locus of a point moves in a plane such that its distance from a fixed point (i.e. focus) is a constant ratio from a fixed line (i.e. directrix). This ratio is called eccentricity and is denoted by 'e'. An ellipse has two foci.

Another definition of an ellipse is the locus of a point which moves in a plane such that the sum of its distance from two fixed points in the same plane is always constant. The parametric representation of an ellipse can be given by Figure 2.2 a generic coordinate of any point on the ellipse.

$$ON = OP_1 cos\beta = acos\beta \tag{2.2}$$

$$PN = P_2Q = OP_2 sin\beta = bsin\beta \tag{2.3}$$

$$P(x,y) \equiv (acos\beta, bsin\beta) \tag{2.4}$$

When the ellipse is rotated about any of the axes, we get a tri-axial ellipsoid. An tri-axial ellipsoid with semi major axis 'a', 'b' and 'c' in the cardinal axis directions 'x', 'y' and 'z' respectively is given by

$$\begin{bmatrix} \frac{1}{a^2} & 0 & 0 \\ 0 & \frac{1}{b^2} & 0 \\ 0 & 0 & \frac{1}{c^2} \end{bmatrix} * \begin{bmatrix} x^2 \\ y^2 \\ z^2 \end{bmatrix} = 1 \tag{2.5}$$

$$\Rightarrow \frac{x^2}{a^2} + \frac{y^2}{b^2} + \frac{z^2}{c^2} = 1 \tag{2.6}$$

If the bi-axial ellipse is rotated around the semi-minor axis 'b', it will generate an ellipsoid of revolution which is also known as an oblate ellipsoid. The mathematical equation of the ellipsoid is given by equation. An ellipsoid is 3D (three dimensional) as depicted in the figure somewhat like a sphere with a uniform bulge at the diameter and depression at the poles. The equation of oblate ellipsoid is given by:

$$\Rightarrow \frac{x^2}{a^2} + \frac{y^2}{a^2} + \frac{z^2}{b^2} = 1 \tag{2.7}$$

The above equation of an ellipsoid can be expressed in matrix form:

$$\begin{bmatrix} \frac{1}{a^2} & 0 & 0 \\ 0 & \frac{1}{a^2} & 0 \\ 0 & 0 & \frac{1}{b^2} \end{bmatrix} * \begin{bmatrix} x^2 \\ y^2 \\ z^2 \end{bmatrix} = 1 \tag{2.8}$$

An auxiliary circle is a circle formed by the locus of the geometric figure when an ellipse is revolved around its minor axis and projected onto a 2D plane as depicted in Figure 2.2.

The dotted circles in the figure are known as auxiliary circles. The line PS is ellipsoid normal at P. The $\angle P^1OA = \beta$ is known as reduced latitude. The $\angle PRA = \theta$ known as geodatic latitude.

Eccentricity and flattening are the computable geometric quantities of an ellipse which gives the sense of its variation from that of a circle. There are

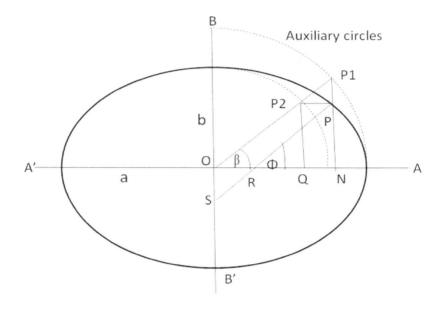

FIGURE 2.2
Auxilary circle, the 2D projected ellipsoid

three variants of eccentricity of ellipse which has a value between $[0,..,1]$. The formulae for first and second eccentricity 'e' of ellipse is given by

$$e^2 = \frac{a^2 - b^2}{a^2} \tag{2.9}$$

$$\Rightarrow a^2 e^2 = a^2 - b^2 \tag{2.10}$$

$$\Rightarrow b^2 = a^2(1 - e^2) \tag{2.11}$$

The secondary eccentricity is denoted by e' and is given by

$$e' = \frac{a^2 - b^2}{b^2} \tag{2.12}$$

Another computable geometric quantity of ellipse is flattening which is given by the formulae first flattening 'f'.

$$f = \frac{a - b}{a} \tag{2.13}$$

The second flattening f' is given by the formula

$$f' = \frac{a - b}{b} \tag{2.14}$$

2.3.1 Relation between 'e' and 'f'

The eccentricity and flattening of any point on the the ellipsoid are related and can be computed given the value of either.

$$f = \frac{a - b}{a} \tag{2.15}$$

Multiplying both sides of the above equation by $(a + b)/a$ we get

$$\frac{(a + b)}{a} f = \frac{(a + b)}{a} \frac{(a - b)}{a} \tag{2.16}$$

$$\frac{(a - b + 2b)}{a} f = e^2 \tag{2.17}$$

$$\frac{a - b}{a} + \frac{2b}{a} f = e^2 \tag{2.18}$$

$$f + 2(1 - f)f = e^2 \tag{2.19}$$

$$2f - f^2 = e^2 \tag{2.20}$$

Therefore if the value of 'e' at any point on the datum surface is given the value of 'f' can be computed and vice versa.

2.4 Computing Radius of Curvature

The radius of curvature at the prime meridian and prime vertical of the ellipsoid are often used for measurement of the grid locations. Radius of curvature at prime meridian is also required to compute Cartesian coordinate (EN) of the corresponding geographic coordinate (latitude, longitude) of any location on the ellipsoid. Also the curvature is an important supplementary parameter for transformation of the geodetic coordinates to Cartesian coordinate or map coordinates.The formulae for radius of curvature of geometric surfaces using differential geometry are given by

$$K = \frac{x' y'' - y' x''}{(x'^2 + y'^2)^{\frac{3}{2}}} \tag{2.21}$$

$$K = \frac{\frac{d^2 y}{dx^2}}{(1 + (\frac{dy}{dx})^2)^{\frac{3}{2}}} \tag{2.22}$$

where, the parametric location of a point on the ellipsoid is given by $x = a\cos\beta$ and $y = b\sin\beta$. Using the above parametric values, their first and second partial derivative can be computed using $x' = -a\sin\beta$ and $x'' = -a\cos\beta$ $y' = b\cos\beta$ and $y'' = -b\sin\beta$.

Therefore, the parametric form of curvature at any point on the ellipsoid can be derived as

$$K = \frac{(-a\sin\beta)(-b\sin\beta) - (-a\cos\beta)(b\cos\beta)}{(a^2\sin^2\beta + b^2\cos^2\beta)^{\frac{3}{2}}} \tag{2.23}$$

$$K = \frac{ab(\sin^2\beta + \cos^2\beta)}{(a^2\sin^2\beta + a^2(1 - e^2)\cos^2\beta)^{\frac{3}{2}}} \tag{2.24}$$

$$K = \frac{b}{a^2(1 - e^2\cos^2\beta)^{\frac{3}{2}}} \tag{2.25}$$

The radius of curvature in parametric form which is inverse of the curvature is given by

$$\rho = \frac{a^2(1 - e^2\cos^2\beta)^{\frac{3}{2}}}{b} \tag{2.26}$$

Given, 'a', 'e' and β the reduced latitude at any point of a datum surface the curvature and radius of curvature can be computed using the above equation. But the reduced latitude is a theoretical quantity. Therefore, one needs to establish a relation ship between the theoretical latitude β and the practical value of the latitude ϕ so that the curvature can be computed for practical purposes. The equation of ellipse is given by

$$\frac{x^2}{a^2} + \frac{y^2}{b^2} = 1 \tag{2.27}$$

On differentiating both sides of the above equation one gets

$$\frac{2x}{a^2} + \frac{2y}{b^2}\frac{dy}{dx} = 0 \tag{2.28}$$

$$\frac{dy}{dx} = -\frac{x}{y}\frac{b^2}{a^2} \tag{2.29}$$

$$\frac{dy}{dx} = -\frac{xa^2(1 - e^2)}{ya^2} = -\frac{x(1 - e^2)}{y} \tag{2.30}$$

where $\frac{dy}{dx}$ is the slope of the tangent at any point $p(x,y)$ of the ellipse. Slope of the normal at the point is negative reciprocal of the slope at the point (by using $m_1m_2 = -1$). Slope of the normal is given by $-\frac{1}{\frac{dy}{dx}}$. By definition, latitude at a point is the slope of the normal at the point. Therefore

$$\tan\phi = \frac{y}{x(1 - e^2)} \tag{2.31}$$

By substituting the parametric value $(a\cos\beta, b\sin\beta)$ of the coordinate (x, y) in the above equation we can obtain the relation.

$$\tan\phi = \frac{a}{b}\tan\beta \tag{2.32}$$

Therefore the radius of curvature at the meridian which was derived in terms of the reduced latitude can be derived for the actual latitude of the point on the ellipsoid for practical use as given below:

$$\rho = \frac{a^2(1 - e^2 cos^2\beta)^{\frac{3}{2}}}{b} \tag{2.33}$$

$$\rho = \frac{a^2}{b}(1 - e^2 \frac{a^2 cos^2\phi}{a^2 cos^2\phi + b^2 sin^2\phi})^{\frac{3}{2}} \tag{2.34}$$

$$\rho = \frac{a^2}{b}(1 - \frac{e^2 a^2 cos^2\phi}{a^2 - e^2 a^2 sin^2\phi})^{\frac{3}{2}} \tag{2.35}$$

$$\rho = \frac{a^2}{b}(1 - \frac{e^2 cos^2\phi}{1 - e^2 sin^2\phi})^{\frac{3}{2}} \tag{2.36}$$

$$\rho = \frac{a^2}{b}(\frac{1 - e^2}{1 - e^2 sin^2\phi})^{\frac{3}{2}} \tag{2.37}$$

$$\rho = \frac{a^2}{a(1 - e^2)^{\frac{1}{2}}}(\frac{1 - e^2}{1 - e^2 sin^2\phi})^{\frac{3}{2}} \tag{2.38}$$

Therefore the radius of curvature at prime meridian given in terms of latitude is

$$\rho = \frac{a(1 - e^2)}{(1 - e^2 sin^2\phi)^{\frac{3}{2}}} \tag{2.39}$$

2.4.1   Radius of Curvature at Prime Vertical Section

Refer to Figure 2.2. The raduis of curvature at the prime vertical section can be derived by using the following equations.

$$PN(radius - of - curvature - at - prime - vertical) = \frac{acos\beta}{cos\phi} \tag{2.40}$$

$$cos\beta = \frac{acos\phi}{cos\phi\sqrt{a^2 cos^2\phi + b^2 sin^2\phi}} \tag{2.41}$$

$$\gamma = \frac{aacos\phi}{cos\phi\sqrt{a^2 cos^2\phi + b^2 sin^2\phi}} \tag{2.42}$$

$$\gamma = \frac{a^2}{\sqrt{a^2 cos^2\phi + a^2(1 - e^2)sin^2\phi}} \tag{2.43}$$

$$\gamma = \frac{a^2}{\sqrt{a^2 - a^2 e^2 sin^2\phi}} \tag{2.44}$$

$$\gamma = \frac{a}{\sqrt{1 - e^2 sin^2\phi}} \tag{2.45}$$

The value of radius of curvatures at $\phi = 0$ i.e. at the equatorial plane and $\phi = 90°$ i.e. at the pole of the ellipsoid one can compute as a special case using the above equations. At $\phi = 0$

$$\rho = a(1 - e^2) = a\frac{b^2}{a^2} = \frac{b^2}{a} \tag{2.46}$$

$$\gamma = a \tag{2.47}$$

$$At\phi = 90^o \tag{2.48}$$

$$\rho = a\frac{1 - e^2}{(1 - e^2)^{\frac{3}{2}}} = \frac{a^2}{b} \tag{2.49}$$

$$\gamma = \frac{a}{\sqrt{1 - e^2}} = \frac{a^2}{b} \tag{2.50}$$

One can observe that the radius curvature at the prime vertical and meridian has the same value.

2.5   Concept of Latitude

We often think of the Earth as simply oval (ellipsoid), but in reality the surface of the Earth is very complex to analyze mathematically. The levels of the surface are different at different position, so for ease of analysis an approximated model of Earth called geoid is used. The geoid is approximated to an ellipsoid (more precisely an oblate ellipsoid). Now, determining the position of a point on the geoid is the real challenge. For this the concept of latitude (angle with respect to equatorial plane) and longitude (angle with respect to the meridian plane) is used. In this section we will discuss the concept of latitude.

2.5.1   Modified Definition of Latitude

Latitude of a point on the geoid is defined as the angle subtended by the normal passing through the point and the equatorial plane. It is usually represented by the symbol ϕ (Greek letter Phi).

There are various different measures of latitude depending upon the model of the geoid. In this section we will discuss the most popular latitude measures often used in geodesy.

2.5.2   Geodetic Latitude

It is the globally accepted definition of latitude. Geodetic latitude of a point on the surface of the geoid is defined as angle subtended by the normal at that point with the equatorial plane (Figure 2.3). It is usually denoted by ϕ. geodetic latitude is taken as the reference for the mathematical definition of the other forms of latitude.

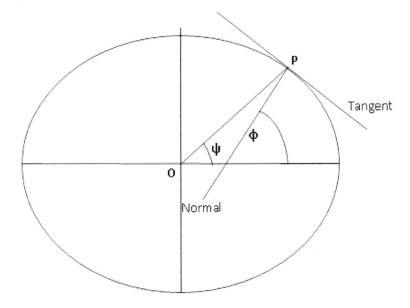

FIGURE 2.3
Geodetic and geocentric latitude

2.5.3   Geocentric Latitude

Geocentric latitude of a point on the geoid is defined as the angle subtended by the radius passing through that point and the equatorial plane. Refer to Figure 2.3. Let 'O' be the center of the geoid and 'P' be an arbitrary point on the surface of the geoid. Therefore the angle subtended by $\overline{OP}$ with the equatorial plane is the Geocentric latitude of that point. It is usually designated by the symbol ψ (Greek small letter Psi). The relation between the geocentric latitude (ψ) and geodetic latitude(ϕ) is given by the equation:

$$\psi(\phi) = \tan^{-1}[(\tan \phi)(1 - e^2)] \tag{2.51}$$

2.5.4   Spherical Latitude

In this case the geoid is assumed to be a perfect sphere. So the normal to any point passes through the center. The angle between that normal and the equatorial plane is called the spherical latitude of that point.

2.5.5   Reduced Latitude

This latitude system is based on the auxiliary circle of the ellipse (more precisely the auxiliary sphere of the ellipsoid) (see Figure 2.4). It is usually de-

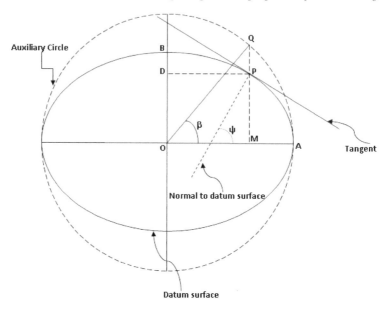

FIGURE 2.4
Reduced latitude

noted by the symbol β (Greek small letter beta). This is also known as **parametric latitude**.

In Figure 2.4, 'O' is the center of the datum surface and 'P' is an arbitrary point on the datum surface. 'M' is the foot of the perpendicular from 'P' on the equatorial plane, similarly 'D' is the foot of the perpendicular from 'P' on the meridian plane.

$PD = OM = q$
$PM = z$
$OB = b$
$OA = OQ = a$

Using geometrical principle

$p = a \cos \beta$
$z = b \sin \beta$

The parametric latitude can be expressed as a function of the geodetic latitude by the equation:

$$\beta(\phi) = \tan^{-1}[\sqrt{1 - e^2} \tan \phi] \qquad (2.52)$$

2.5.6 Rectifying Latitude

The concept of rectifying latitude is based on meridian distance. On a sphere the normal passes through the center and the latitude is therefore equal to the angle subtended at the prime meridian arc from the equator to the point concerned. The meridian distance of the point is given by:

$$m(\phi) = R.\phi \tag{2.53}$$

In the case of rectifying latitude the meridian distance is scaled so that the values of the poles is $\frac{\pi}{2}$. Rectifying latitude is denoted by the symbol 'μ' (Greek small letter mu) and can be expressed as a function of Geodetic latitude:

$\mu(\phi) = \frac{\pi}{2} \frac{m(\phi)}{m_p}$

where, $m(\phi)$ = meridian distance from equator to latitude ϕ, and is given by:

$a(1 - e^2) \int\limits_{0}^{\phi} (1 - e^2 \sin^2 \phi)^{\frac{3}{2}} d\phi$

m_p is the length of meridian quadrant from equator to pole.

$m_p = m(\frac{\pi}{2})$

From the rectifying latitude approximation, the radius of the geoid is given by:

$$R_\mu = \frac{2m_p}{\pi} \tag{2.54}$$

2.5.7   Authalic Latitude

Authalic latitude gives an area preserving transformation to a sphere. It is usually denoted by the symbol ξ (Greek small letter xi). Albers equal area conic projection is an example of the use of authalic latitude. It can be represented as a function of geodetic latitude by the equation:

$$\xi(\phi) = \sin^{-1} \frac{q(\phi)}{q_p} \tag{2.55}$$

where:

$q(\phi) = \frac{(1-e^2)\sin\phi}{1-e^2\sin^2(\phi)} - \frac{1-e^2}{2e} \ln \frac{1-e\sin\phi}{1+e\sin\phi}$

$= \frac{(1-e^2)\sin\phi}{1-e^2\sin^2(\phi)} - \frac{1-e^2}{e} \sinh^{-1}(e\sin\phi)$

$q_p = q(\frac{\pi}{2}) = 1 - \frac{1-e^2}{2e} \ln(\frac{1-e}{1+e})$

$= 1 + \frac{1-e^2}{e} \tanh^{-1} e$

From the authalic approximation, the radius of the geoid is given by:

$$R_q = a\sqrt{\frac{q_p}{2}} \tag{2.56}$$

2.5.8   Conformal Latitude

The conformal latitude defines a transformation from the ellipsoid to a sphere of arbitrary radius such that the angle of intersection between any two lines on the ellipsoid is the same as the corresponding angle on the sphere. That means it retains the conformality of mapping by giving an area preserving transformation. It is usually denoted by the symbol χ (Greek small letter chi) and can be expressed as a function of geodetic latitude by the equation:

$$\chi(\phi) = 2\tan^{-1}\left[\left(\frac{1+\sin\phi}{1-\sin\phi}\right)\left(\frac{1-e\sin\phi}{1+e\sin\phi}\right)^e\right]^{\frac{1}{2}} - \frac{\pi}{2} \qquad (2.57)$$

$$\Rightarrow \chi(\phi) = \sin^{-1}[\tanh(\tan^{-1}(\sin\phi)) - e\tanh^{-1}(e\sin\phi)] \qquad (2.58)$$

2.5.9   Isometric Latitude

Isometric latitude is used in development of normal Mercator projection and transverse Mercator projection. This is usually denoted by the symbol Ψ (Greek capital letter Psi - not to be confused with geocentric latitude). The name isometric arises from the fact that at any point on the ellipsoid, equal increments of Ψ and λ gives equal distance displacements along the meridian and parallels respectively. The graticules defined by the lines of constant Ψ and constant λ divide the surface of the ellipsoid into a mesh of squares (of variable size). The isometric latitude is 0 at the equator, but rapidly diverges from geodetic latitude and becomes infinite at poles. It can be expressed as a function of geodetic latitude by the equation:

$$\Psi(\phi) = \ln[\tan(\frac{\pi}{4} + \frac{\phi}{2})] + \frac{e}{2}\ln[\frac{1-e\sin\phi}{1+e\sin\phi}] \qquad (2.59)$$

$$\Rightarrow \Psi(\phi) = \tanh^{-1}(\sin\phi) - e\tan^{-1}(e\sin\phi) \qquad (2.60)$$

2.5.10   Astronomical Latitude

This is the angle between the equatorial plane and the true vertical at a point on the surface. The true vertical is the direction of gravity field at that point (the gravity field is the resultant of the acceleration due to gravity and centrifugal acceleration at that point).

Astronomical latitude is calculated from angles measured between the zenith and the stars whose declination is actually known.

The zenith is an imaginary point directly above a particular location on the celestial field. Above means in the vertical direction opposite to the apparent gravitational force at that point.

2.6 Applications of Geodesy

Geodesic methods find many applications such as engineering survey, precisely locating the position of Earth objects by triangulation, trilateration and traverse. Some of the important applications of geodesy are as follows.

1. Establishment of triangulation for geodatic survey by triangulation, trilateration and traverse.

2. Measurement of height of objects and locations above the sea level by triangulation and spirit leveling.

3. Computation of latitude, longitude and azimuth of locations through astronomical observations.

4. Observation of direction of gravity by astronomical observations of latitude, longitude.

5. Determination of the relative change in position of ground and its height from sea level for determination of crustal movement.

Some of the geodesic applications are:

1. To study the polar motion and its effect on the location of objects.

2. Primary or zero order triangulation, trilateration and traverse.

3. Astronomical observation of latitude, longitude and azimuth to locate origins of surveys, and to control their direction.

4. Observation of the intensity of gravity by the pendulum and other apparatus.

5. To deduce the exact form of Earth's sea level equi-potential surfaces at all heights.

6. Polar motion studies.

7. Earth tides.

8. The separation between the geoid and the mean sea level.

9. Engineering survey.

10. Satellite geodesy: includes the modern techniques of positioning by space method. e.g. GPS, SLR, VLBI, etc.

2.7 The Indian Geodetic Reference System (IGRS)

The Indian geodetic reference system has taken into account the existence of Mount Everest in its region. Hence the ellipsoidal model of the Indian region has computed a model known as the Everest ellipsoid. The constants of the Everest ellipsoid were determined in 1830 and are given by:

1. a (semi major axis) = 6377301.243 m

2. b (semi major axis) = 6356100.231 m

3. f (flattening) = $\frac{a-b}{a}$ = $\frac{1}{300.8017}$

4. e^2 (square of eccentricity) = $\frac{(a^2-b^2)}{a^2}$ = 0.00663784607

2.8   Summary

Geodesic computations play a major role in computing the geometric quantities of the Earth. Geodesy models the shape of Earth through observations of various physical phenomena exhibited by the Earth such as its gravity field, rotational motion etc. This chapter starts with the definition of geodesy followed with the definition of the mathematical models of Earth such as geoid and ellipsoid. The geometric quantities describing the ellipsoid such as eccentricity, flattening are derived. The computation of radius of curvature prime vertical and meridian are derived. The chapter gives a elaborate definition and mathematical formulations of different types of latitudes such as geodetic, geocentric, reduced, authalic, conformal, astronomical latitude and their inter relationships are given with mathematical formulations so as to compute them from geodetic latitude. Applications of geodesy and its allied computing methods are in use for measurement of different physical phenomena exhibited by Earth.

3

Reference Systems and Coordinate Transformations

If geodesy is about modelling the Earth's surface and shape, the coordinate system connects the model produced by geodesic modelling to a frame of reference for measurement. The coordinate system or reference frame or frame of reference is a mathematical concept for imparting order to a distributed set of objects in sea, space or terrain. The frame of reference helps in conceptualizing the location of the objects with respect to the common frame and with respect to its surroundings objects. A reference system is a mathematical concept for modelling the real world. This chapter starts with the mathematical definition of a coordinate reference system. Then we discuss different types of coordinate systems and their characterization, since the mathematical and computing aspects of GIS used to represent a point in the coordinate system are quite intriguing. Map projection and coordinate system are interrelated mathematically. A discussion on coordinate systems and map projections can be found in Maling [35].

3.1 Definition of Reference System

A reference system is defined by the following:

1. Origin of the reference system

2. The orthogonal or non-orthogonal directrix as reference frames with orientations with respect to each other.

The dimension of the reference system is the number of directrix used to describe the objects. Therefore depending on the number of directrix in the reference frame the coordinate system can be classified as 1D,2D,3D,..,nD, where 'D' stands for dimension. Depending on the orientation of the coordinates with respect to each other they can be classified as orthogonal or non-orthogonal.

The origin of the frame of reference can be fitted to a real world object such as planet Earth, the moon, a constellation of stars, space, the center of a city or a building etc. Also it can be attached to some known geometrical shape such as a sphere, cylinder or plane. Depending on the position of the origin and its orientation with geometrical object the coordinate system can be classified into world coordinate system, Earth Centered Earth Fixed coordinate system (ECEF), celestial coordinate system. According to geometrical shape reference frame can be classified into planar, cylindrical, spherical etc. coordinate reference systems are useful in

1. Imparting an orderliness to a set of spatially distributed objects.

2. Imparting topological ordering to spatial objects in the frame of reference.

3. As a concept for mathematical method for geometric and spatial modelling.

Hence coordinate systems are handy and useful in geometry modelling, ordering, indexing visualization and computations of spatio-temporal objects. Therefore the coordinate reference system is integral part of GIS in general and of spatio-temporal systems in particular.

3.2   Classification of Reference Systems

There are a number of coordinate systems in use in mathematics, geodesy, GIS and computer graphics. Classifying them into few major groups is very difficult because many of them possess similar mathematical properties classifying them into more than one type of system. The major classification criteria for coordinate systems can be as follows.

1. Depending upon the dimension or number of axis of the coordinate system: 1D,2D,3D,..,nD

2. Depending on the orientation of the axis: orthogonal and non-orthogonal

3. Depending upon the quantity for representation of the coordinate : rectangular coordinate system also known as Cartesian coordinate system or curvilinear (polar) coordinate system

4. Depending upon the orientation of the coordinate system: right handed coordinate system or left handed coordinate system

5. Depending upon the area of influence or extend of representation: local coordinate system, global coordinate system or universal coordinate system

6. Depending upon the object to which the coordinate system is embedded: planar, spherical, conical, cylindrical coordinate system

7. Miscellaneous category of coordinate systems are homogenous coordinate system or non homogenous coordinate system and Barycentric coordinate system

3.3 Datum and Coordinate System

The foundation of a coordinate system is a datum surface. Coordinates without a specified datum have no meaning because they cannot answer questions such as:

1. Where is the origin of the coordinate system?
2. What is the height of the coordinate of a point? Or height above what reference plane?
3. On which or what surface is the object located or lies?

A meaningful real world coordinate system must answer these questions unambiguously. If it cannot answer such questions there is no real world attachment and real use of the coordinates.

3.4 Attachment of Datum to the Real World

A datum surface or a datum model attaches a real world system to a hypothetical frame of reference making it meaningful. Therefore, for a coordinate system to become meaningful a numerical origin, or a starting point, is a necessity. Also a clearly defined surface is necessary for the real world spatial objects to correlate it with the theoretical frame of reference. A datum can be thought of as a foundation to the otherwise abstract coordinate system. Also datum can be thought of as an architectural drawing of a building to be constructed. Therefore attachment of an abstract datum with a real world object is essential for meaningful computations.

Datum is a mathematically modelled surface corresponding to a real world object such as Earth. Therefore datum surfaces exhibit perfect mathematical and geometrical behaviour and are abstract errorless surfaces. On a datum every point has a unique and accurate coordinate. There is no distortion and ambiguity of location. For example, the position of any point on the datum can be stated exactly, and it can be accurately transformed into coordinates on another datum with no discrepancy whatsoever. All of these wonderful things are possible only as long as a datum has no connection to anything in the physical world. In that case, it is perfectly accurate and perfectly useless.

Suppose, however, that one wishes to assign coordinates to objects on the floor of a very real rectangular room. A Cartesian coordinate system could work, if it is fixed to the room with a well-defined orientation. For example, one could put the origin at the southwest corner and use the floor as the reference plane. With this datum, one not only has the advantage that all of the coordinates are positive, but one can also define the location of any object on the floor of the room. The coordinate pairs would consist of two distances, the distance east and the distance north from the origin in the corner. As long as everything stays on the floor, you can assign numerical pairs of coordinates to the object. In this case, there is no error in the datum, of course, but there are inevitably errors in the coordinates. These errors are due to the less-than-perfect flatness of the floor, the impossibility of perfect measurement from the origin to any object, the ambiguity of finding the precise center of any of the objects being assigned coordinates, and similar factors. In short, as soon as one brings in the real world, things get messy. Now if the rectangular room is a part of a multi storied building than the origin can be attached to the base of the building with altitude as the third dimension in addition to east and north to describe the aerial location of the rooms in the building.

3.5 Different Coordinate Systems Used in GIS

Coordinate systems are a key characteristic of spatial data. They give spatial ordering to spatial objects for searching, sorting, visualizing and understanding the spatial relationship of an object with its surroundings. The spatial ordering of spatial objects helps to understand and create a mental picture of the terrain. There are many well known coordinate systems being used for representation of spatial data. Each of these coordinate systems has some mathematical properties which make it suitable for representing spatial data associated with a particular domain. This section describes the key features of different coordinate systems. The concept of datum or geo-datic datum to mathematically model the shape of the Earth is important before applying a coordinate system. The concept of datum with specific emphasis on WGS-84 (World Geodatic Datum 1984) which is a terrestrial reference frame universally accepted by many agencies worldwide is discussed.

There are many coordinate systems prevailing and under use in GIS. Some prominent coordinate system used in GIS are listed below. Each one of these coordinate systems has its strengths and weaknesses and was developed with different constraints keeping in mind the specific application. Therefore the application specific coordinate systems are:

1. Rectangular Cartesian coordinate system

2. Geographic coordinate system

3. Spherical coordinate system

4. Cylindrical coordinate system

5. Polar and log-polar coordinate system

6. Earth centered Earth fixed (ECEF) coordinate system

7. Inertial terrestrial reference frame (ITRF)

8. Concept of grid, UTM, Mercator's grid and military GRID

9. Celestial coordinate system

3.5.1 The Rectangular Coordinate System

Cartesian coordinates then are rectangular, or orthogonal if one prefers, defined by perpendicular axes from an origin, along a reference surface. These elements can define a datum, or framework, for meaningful coordinates. As a matter of fact, two-dimensional Cartesian coordinates are an important element in the vast majority of coordinate systems, State plane coordinates in the United States, the Universal Transverse Mercator (UTM) coordinate system, and most others. The datums for these coordinate systems are well established. There are also local Cartesian coordinate systems whose origins are often entirely arbitrary. For example, if surveying, mapping, or other work is done for the construction of a new building, there may be no reason for the coordinates used to have any fixed relation to any other coordinate systems. In that case, a local datum may be chosen for the specific project with north and east fairly well defined and the origin moved far to the west and south of the project to ensure that there will be no negative coordinates. Such an arrangement is good for local work, but it does preclude any easy combination of such small independent systems. Large scale Cartesian datums, on the other hand, are designed to include positions across significant portions of the Earth's surface into one system. Of course, these are also designed to represent our decidedly round planet on the flat Cartesian plane, which is no easy task. But how would a flat Cartesian datum with two axes represent the Earth? Distortion is obviously inherent in the idea.

3.5.2 The Spherical Coordinate System

A spherical coordinate system is a coordinate system for three-dimensional space where the position of a point is specified by three numbers: (a) the radial distance of that point from a fixed origin, (b) its polar angle measured from a fixed zenith direction, (c) the azimuth angle of its orthogonal projection on a reference plane that passes through the origin and is orthogonal to the zenith, measured from a fixed reference direction on that plane.

To define a spherical coordinate system, one must choose two orthogonal directions, the zenith and the azimuth reference, and an origin point in space.

These choices determine the reference plane that contains the origin and is perpendicular to the zenith. The spherical coordinates of a point P are then defined as follows:

1. The radius or radial distance 'r' which is the Euclidean distance from the origin O to P.

2. The inclination θ (or polar angle) is the angle between the zenith direction and the line segment OP.

3. The azimuth ϕ (or azimuthal angle) is the signed angle measured from the azimuth reference direction to the orthogonal projection of the line segment OP on the reference plane.

4. The sign of the azimuth is determined by choosing what is a positive sense of turning about the zenith. This choice is arbitrary, and is part of the coordinate system's definition.

Often the radial distance is also called the radius or radial coordinate. The polar angle may be called co-latitude, zenith angle, normal angle, or inclination angle. The use of symbols and the order of the coordinates differ for different areas of applications. Generally these quantities are given by the triple (r, θ, ϕ) or (ρ, θ, ϕ) which is interpreted as radial distance, polar angle, and azimuthal angle in physics where as in mathematics it is interpreted as radial distance, azimuthal angle, and polar angle.

In GIS the spherical system is known as the geographic coordinate system where the location of the spatial objects are expressed by (latitude, longitude), where the radial distance which is equivalent to the mean radius of the Earth is implicit and is treated as a constant for the spherical datum. The latitude is the complement of the polar angle and the longitude is equivalent to the azimuthal angle of the corresponding conventional spherical system.

There are a number of different celestial coordinate systems based on different fundamental planes and with different terms for the various coordinates. The spherical coordinate systems used in mathematics normally use radians rather than degrees and measure the azimuthal angle counter clockwise rather than clockwise. The inclination angle is often replaced by the elevation angle measured from the reference plane. Elevation angle of zero is known as the horizon. The concept of spherical coordinates can be extended to higher dimensional spaces and is then referred to as hyper-spherical coordinates.

3.5.3   The Cylindrical Coordinate System

A cylindrical coordinate system is a 3D reference frame which specifies each point in the reference frame by a 3-triple (ρ, Φ, z), where:

$\rho = $ the distance of the point from the cylindrical axis of the reference frame known as the radial distance or radius.

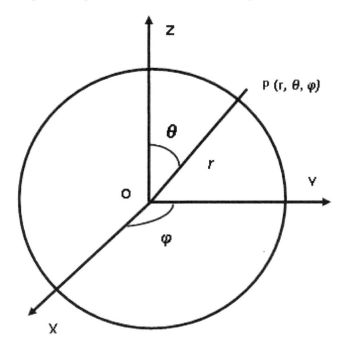

FIGURE 3.1
Spherical coordinate system

Φ = the azimuth angle between the reference direction on the chosen plane and the line from the origin to the projection of 'ρ' on the plane.

z = the height, which is a signed distance from the chosen plane to the point P.

The reference system consists of an axis which is symmetrically placed to the center of a cylindrical space or object as depicted in Figure 3.2. The points in the space are specified by distance from the chosen axis, the direction from the axis relative to the reference direction and the reference plane perpendicular to the axis. Often the axis is referred as the cylindrical axis so that it can be unambiguously referred axis of cylindrical reference frame rather than the polar axis. The cylindrical reference system is embedded to objects which have cylindrical shape or symmetry along the longitudinal axis.

The cylindrical coordinate system is useful for projection of Earth's datum surface. It is used to describe the objects and phenomena which have rotational symmetry about the longitudinal axis such as the electromagnetic field produced by a cylindrical conductor carrying current or to position the stars in a galaxy.

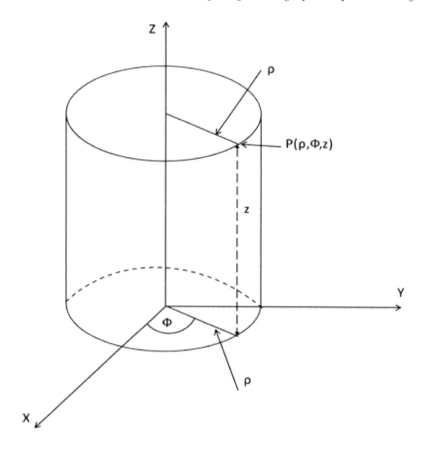

FIGURE 3.2
Cylindrical coordinate system

3.5.4   The Polar and Log-Polar Coordinate System

There is another way of looking at a direction. It can be one component
of a coordinate. A procedure familiar to surveyors using optical instruments
involves the occupation of a station with an established coordinate. A back
sighting is taken either on another station with a coordinate on the same
datum or some other reference, such as Polaris. With two known positions, the
occupied and the sighted, a beginning azimuth or bearing is calculated. Next,
a new station is sighted ahead, or fore-sighted, on which a new coordinate will
be established. The angle is measured from the back sight to the fore sight,
fixing the azimuth or bearing from the occupied station to the new station. A
distance is measured to the new station. This direction and distance together
can be considered the coordinate of the new station. They constitute what is

known as a polar coordinate. In surveying, polar coordinates are often a first step toward calculating coordinates in other systems.

A polar coordinate defines a position with an angle and distance. As in a Cartesian coordinate system, they are reckoned from an origin, which in this case is also known as the center or the pole. The angle used to define the direction is measured from the polar axis, which is a fixed line pointing to the east, in the configuration used by mathematicians. Note that many disciplines presume east as the reference line for directions, CAD utilities, for example. Mappers, cartographers, and surveyors tend to use north as the reference for directions in polar coordinates.

In the typical format for recording polar coordinates, the Greek letter rho, ρ, indicates the length of the radius vector, which is the line from the origin to the point of interest. The angle from the polar axis to the radius vector is represented by the Greek letter theta, θ, and is called the vectorial angle, the central angle, or the polar angle. These values, ρ and θ, are given in ordered pairs, like Cartesian coordinates. The length of the radius vector is first and the vectorial angle second - for example, $P_1(65, 6450), P_2(77, 7292)$.

There is a significant difference between Cartesian coordinates and polar coordinates. In an established datum using Cartesian coordinates, one and only one ordered pair can represent a particular position. Any change in either the northing or the easting implies that the coordinate represent a completely different point. However, in the polar coordinates the same position might be represented in many different ways, with many different ordered pairs standing for the very same point. For example, (65, 45) can just as correctly be written as (65, 405). Here the vectorial angle swings through 360 degrees and continues past the pole through another 45 degrees. It could also be written as (65, -315). In other words, there are several ways to represent the same point in polar coordinates. This is not the case in rectangular coordinates, nor is it the case for the polar coordinate system used for surveying, mapping, and cartography. In mapping and cartography, directions are consistently measured from north and the polar axis points north as shown in Figure 3.3. For mathematical computations and mathematical convenience in the arrangement of polar coordinates, a counter clockwise vector angle is treated as negative and a clockwise measured angle is positive. The angle may be measured in degrees, radians, or grads, but if it is clockwise, it is positive.

3.5.5 Earth-Centered Earth-Fixed (ECEF) Coordinate System

The ECEF coordinate system is a right handed, 3D, barycentric coordinate system where the center of the coordinate system is fixed at the center of mass of Earth. The positive Z-axis is along the rotation axis of the Earth, X-Y plane is the equatorial plane of the Earth with the X-axis perpendicular to the prime meridian (zero point of longitude). The Y-axis lies in the equatorial plane perpendicular to the X-axis and set to make it right handed (it is directed

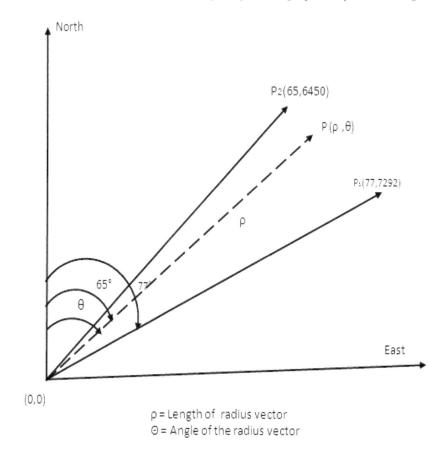

p = Length of radius vector
Θ = Angle of the radius vector

FIGURE 3.3
Polar coordinate system

towards the Indian Ocean). Since Earth is a non rigid body with varying
mass density, the center of mass of Earth is computed by solving the volume
integral given by the integral summation of position vector times the density
vector equated to zero.

$$\sum \rho(x, y, z) = 0 \qquad (3.1)$$

Since the center of mass of Earth is varying due to numerous disturbing forces
therefore the center of the ECEF is also varying. Also the axis of rotation of
Earth is in a non uniform rotational axis. Rather it is wobbling due to many
non secular gravitational forces. Hence though the ECEF is known as Earth
centered and Earth fixed, nothing is fixed in this coordinate system, rather
it is constantly moving and rotating along with the Earth. ECEF is used by
most of the satellite sensors and Radars to fix the location of ground objects
or objects in space. Most of the Global Positioning Systems (GPS) use ECEF

to specify the location of objects on the surface or subsurface of the Earth. ECEF is used extensively because it does not need any datum parameter to fix the position of objects. It only needs the center of mass of Earth and the orientation parameter of its axis. To convert a coordinate given in ECEF to any other Earth coordinate the geodetic datum parameter is necessary as an additional parameter. The ECEF coordinate system is essentially a 3D Cartesian coordinate system.

3.5.6 Inertial Terrestrial Reference Frame (ITRF)

The Earth is constantly changing shape. To understand the context, when the motion of the Earth's crust is observed, it must be referenced. A terrestrial reference frame provides a set of coordinates of some points located on the Earth's surface. It can be used to measure plate tectonic and regional subsidence. It is also used to represent the Earth when measuring its rotation in space. This rotation is measured with respect to a frame tied to stellar objects, called a celestial reference frame. The International Earth Rotation and Reference Systems Service (IERS) was created in 1988 to establish and maintain a celestial reference frame, the ICRF, which is a terrestrial reference frame, the ITRF. The Earth Orientation Parameters (EOPs) connect these two frames. These frames provide a common reference to compare observations and results from different locations. Nowadays, four main geodetic techniques are used to compute accurate coordinates: the GPS, VLBI, SLR, and DORIS. Since the tracking network equipped with the instruments of those techniques is evolving and the period of data available increases with time, the ITRF is constantly being updated. There were eleven realizations of the ITRS were set up from 1988. The latest is the ITRF2005. All these realizations include station positions and velocities. They model secular Earth's crust changes which is why they can be used to compare observations from different epochs. All the higher frequencies of the station displacements can be accessed with the IERS conventions. Continuity between the realizations has been ensured as much as possible when adopting conventions for ITRF definitions. The relationship linking all these solutions is of utmost importance. They are supplied here by the transformation parameters.

The International Terrestrial Reference System (ITRS) is a world spatial reference system co-rotating with the Earth in its diurnal motion in space. The IERS, in charge of providing global references to the astronomical, geodetic and geophysical communities, supervises the realization of the ITRS. Realizations of the ITRS are produced by the IERS ITRS Product Center (ITRS-PC) under the name International Terrestrial Reference Frames (ITRF). ITRF coordinates were obtained by combination of individual TRF solutions computed by IERS analysis centers using the observations of Space Geodesy techniques: GPS, VLBI, SLR, LLR and DORIS. They all use networks of stations located on sites covering the whole Earth.

A Terrestrial Reference System (TRS) is a spatial reference system co-rotating with the Earth in its diurnal motion in space. In such a system, positions of points anchored on the Earth's solid surface have coordinates which undergo only small variations with time, due to geophysical effects (tectonic or tidal deformations). A Terrestrial Reference Frame (TRF) is a set of physical points with precisely determined coordinates in a specific coordinate system (Cartesian, geographic, mapping) attached to a TRS. Such a TRF is said to be a realization of the TRS.

An ideal TRS is defined as a tri-dimensional reference frame (in the mathematical sense) close to the Earth and co-rotating with it. In the Newtonian background, the geometry of the physical space considered as an euclidian affine space of three dimension provides a standard and rigorous model of such a system through the selection of an affine frame (O,E). O is a point of the space named origin. E is a vector base of the associated vector space.The currently adopted restrictions to E are to be orthogonal with same length for the base vectors. Moreover, one adopts a direct orientation. The common length of these vectors will express the scale of the TRS and the set of unit vectors collinear to the base of its orientation.

3.5.7 Celestial Coordinate System

It is useful to impose on the celestial sphere a coordinate system that is analogous to the latitude-longitude system employed for the surface of the Earth. This model is also known as astronomy without a telescope. Such a coordinate system is illustrated in the Figure 3.4. In celestial coordinate system the Earth should be imagined to be a point at the center of the celestial sphere.

In the celestial coordinate system the north and south celestial poles are determined by projecting the rotation axis of the Earth to intersect the celestial sphere, which in turn defines a celestial equator. The celestial equivalent of latitude is called declination and is measured in degrees north (positive numbers) or south (negative numbers) of the celestial equator. The celestial equivalent of longitude is called right ascension. Right ascension can be measured in degrees, but for historical reasons it is more common to measure it in time (hours, minutes, seconds): the sky turns 360 degrees in 24 hours and therefore it must turn 15 degrees every hour; thus, 1 hour of right ascension is equivalent to 15 degrees of (apparent) sky rotation.

The zero point for celestial longitude (that is, for right ascension) is the vernal equinox, which is that intersection of the ecliptic and the celestial equator near where the Sun is located in the Northern Hemisphere Spring. The other intersection of the celestial equator and the ecliptic is termed the autumnal equinox. When the Sun is at one of the equinoxes the lengths of day and night are equivalent (equinox derives from a root meaning 'equal night'). The time of the Vernal Equinox is typically about March 21 and of the autumnal equinox about September 22. The point on the ecliptic where the Sun is most north of the celestial equator is termed the summer solstice and the point

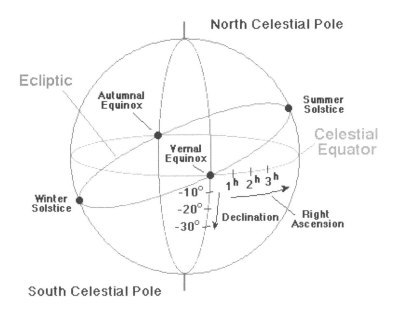

FIGURE 3.4
Celestial coordinate system

where it is most south of the celestial equator is termed the winter solstice. In the northern hemisphere the hours of daylight are longest when the Sun is near the summer solstice (around June 22) and shortest when the Sun is near the winter solstice (around December 22). The opposite is true in the southern hemisphere. The term solstice derives from a root that means to 'stand still'; at the solstices the Sun reaches its most northern or most southern position in the sky and begins to move back toward the celestial equator. Thus, it 'stands still' with respect to its apparent north-south drift on the celestial sphere at that time. Traditionally, northern hemisphere spring and autumn begin at the times of the corresponding equinoxes, while northern hemisphere winter and summer begin at the corresponding solstices. In the southern hemisphere, the seasons are reversed (e.g., southern hemisphere spring begins at the time of the autumnal equinox).

The right ascension (RA) and declination (dec) of an object on the celestial sphere specify its position uniquely, just as the latitude and longitude of an object on the Earth's surface define a unique location. Thus, for example, the star Sirius has celestial coordinates 6 hr 45 min RA and -16 degrees 43 minutes declination, as illustrated in the Figure This tells us that when the vernal equinox is on our celestial meridian, it will be 6 hours and 45 minutes before Sirius crosses our celestial meridian, and also that Sirius is a little more than 16 degrees south of the celestial equator.

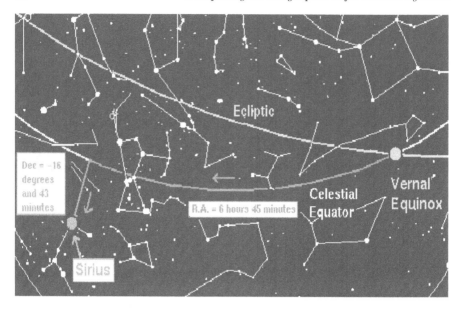

FIGURE 3.5
Celestial coordinate of constellation Sirus defined by RA and declination

3.5.8   Concept of GRID, UTM, Mercator's GRID and Military GRID

Within each zone we draw a transverse Mercator projection centered on the middle of the zone. Thus for zone 1, with longitudes ranging from 180 degrees west to 174 degrees west, the central meridian for the transverse Mercator projection is 177 degrees west. Since the equator meets the central meridian of the system at right angles, we use this point to orient the grid system as depicted in Figure 3.6. Two forms of the UTM system are in common use. The first, used for civilian applications, sets up a single grid for each zone. To establish an origin for the zone, we work separately for the two hemispheres. For the southern hemisphere, the zero northing is the south pole, and we give northings in meters north of this reference point. Fortunately, the meter was originally defined as one ten millionth of the distance from the pole to the equator, actually measured on the meridian passing through Paris. While the distance varies according to which meridian is measured, the value 10 million is sufficient for most cartographic applications.

The numbering of northings start again at the equator, which is either 10,000,000 meters north in southern hemisphere coordinates or '0' meters north in northern hemisphere coordinates. Northings then increase to 10,000,000 meters at the North Pole Note that as we approach the poles the distortions of the latitude-longitude grid drift farther and farther from the

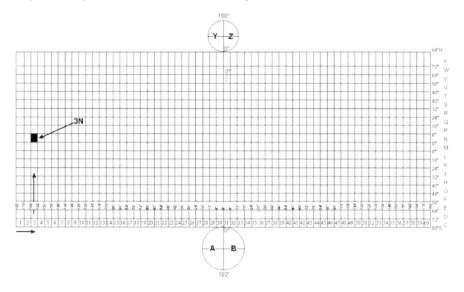

FIGURE 3.6
Universal transverse Mercator grid system

UTM grid. It is customary, therefore, to use the UTM system neither beyond the land limits of North America, nor for the continent of Antarctica. This means that the limits are 84 degrees north and 80 degrees south. For the polar regions, the Universal Polar Stereographic (UPS) coordinate system is used.

For eastings a false origin is established beyond the westerly limit of each zone. The actual distance is about half a degree, but the numbering is chosen so that the central meridian has an easting of 500,000 meters. This has the dual advantage of allowing overlap between zones for mapping purposes, and of giving all eastings positive numbers. Also, we can tell from our easting if we are east or west of the central meridian, and therefore the relationship between true north and grid north at any point. To give a specific example, Hunter College is located at UTM coordinate 4,513,410 meters north; 587,310 meters east; zone 18, northern hemisphere. The reader is advised to locate Hunter College on a Google image or WikiMapia. UTM grid north would therefore appear to be east of true north. Another example of expressing location through UTM grid coordinate is the location of Sir C.V. Raman Nagar, Bangalore in India which has UTM coordinate 143637 meters north; 788804 meters east, in the UTM zone 43P at a height of 900 meters from MSL. The geographic reference of Sir C.V. Raman Nagar is expressed by $12° : 59' : 44''$ latitude and $77° : 39' : 43''$ longitude and falls in the Survey of India's 1:50K topo-sheet number 57H9. The reader is advised to locate Hunter College and Sir C.V. Raman Nagar on a Google image or WikiMapia. UTM grid north would therefore appear to be east of true north.

For geo-coding of objects using the UTM system 16 digits are enough to store the location to a precision of 1 meter, with one digit restricted to a binary (northern or southern hemisphere), and the first digit of the zone restricted to 0 to 6 (60 is the largest zone number).

This coordinate system has two real cartographic advantages. First, geometric computations can be performed on geographic data as if they were located not on the surface of a sphere but on a plane. Over small distances, the errors in doing so are minimal, although it should be noted that area computations over large regions are especially cartographically dangerous. Distances and bearings can similarly be computed over small areas. The second advantage is that the level of precision can be adapted to the application. For many purposes, especially at small scales, the last UTM digit can be dropped, decreasing the resolution to 10 meters. This strategy is often used at scales of 1:250,000 and smaller. Similarly, sub meter resolution can be added simply by using decimals in the eastings and northings. In practice, few applications except for precision surveying and geodesy need precision of less than 1 meter, although it is often used to prevent computer rounding error.

3.6　Shape of Earth

People have been proposing theories about the shape and size of the planet Earth for thousands of years. In 200 B.C. Eratosthenes got the circumference about right, but a real breakthrough came in 1687 when Sir Isaac Newton suggested that the Earth's shape was ellipsoidal in the first edition of his *Principia*. The shape of Earth is better described through a grid of latitudes and longitudes.

3.6.1　Latitude and Longitude

Latitude and longitude are coordinates that represent a position with angles instead of distances. Usually the angles are measured in degrees, but grads and radians are also used. Depending on the precision required, the degrees (with 360 degrees comprising a full circle) can be subdivided into 60 minutes of arc, which are themselves divided into 60 seconds of arc. In other words, there are 3600 sec in a degree. Seconds can be subsequently divided into decimals of seconds. The arc is usually dropped from their names, because it is usually obvious that the minutes and seconds are in space rather than time. In any case, these subdivisions are symbolized by θ for degrees, $'$, for minutes, and $''$ for seconds. This system is called sexadesimal.

Lines of latitude and longitude always cross each other at right angles, just like the lines of a Cartesian grid, but latitude and longitude exist on a curved rather than a flat surface. There is imagined to be an infinite number of these lines on the ellipsoidal model of the Earth. In other words, any and every place has a line of latitude and a line of longitude passing through it, and it takes both of them to fully define a place. If the distance from the surface of the ellipsoid is then added to a latitude and a longitude, one has a three-dimensional (3D) coordinate. This distance component is sometimes the elevation above the ellipsoid, also known as the ellipsoidal height, and sometimes it is measured all the way from the center of the ellipsoid.

In mapping, latitude is usually represented by ϕ (the Greek small letter phi), longitude is usually represented by λ (the Greek small letter lambda). In both cases the angles originate at a plane that is imagined to intersect the ellipsoid. In both latitude and longitude, the planes of origination, are intended to include the center of the Earth. Angles of latitude most often originate at the plane of the equator, and angles of longitude originate at the plane through an arbitrarily chosen place, now Greenwich, England. Latitude is an angular measurement of the distance a particular point lies north or south of the plane through the equator measured in degrees, minutes, seconds, and usually decimals of a second. Longitude is also an angle measured in degrees, minutes, seconds, and decimals of a second east and west of the plane through the chosen prime, or zero, position.

3.6.2 Latitude

Two angles are sufficient to specify any location on a reference ellipsoid representing the Earth. Latitude is an angle between a plane and a line through a point.

Imagine a flat plane intersecting an ellipsoidal model of the Earth. Depending on exactly how it is done, the resulting intersection would be either a circle or an ellipse, but if the plane is coincident or parallel with the equator, as all latitudes are, the result is always a parallel with the latitude. The equator is a unique parallel of latitude that also contains the center of the ellipsoid. A flat plane parallel to the equator creates a circle of latitude smaller then the equator.

The equator is 0° latitude, and the North and south poles have +90° north and −90° south latitude respectively. In other words, values for latitude range from a minimum of 0° to a maximum of 90°. The latitudes north of the equator are positive, and those to the south are negative.

Lines of latitude, circles actually, are called parallels because they are always parallel to each other as they proceed around the globe. They do not converge as meridians do or cross each other.

3.6.3 Longitude

Longitude is an angle between two places. In other words, it is a dihedral angle. A dihedral angle is measured at the intersection of the two planes.

The first plane passes through the point of interest, and the second plane passes through an arbitrarily chosen point agreed upon as representing zero longitude. That place is Greenwich, England. The measurement of angles of longitude is imagined to take place where the two planes meet, the polar axis that is the axis of rotation of the ellipsoid.

These planes are perpendicular to the equator, and where they intersect the ellipsoidal model of the Earth they create an elliptical line on its surface. The elliptical line is then divided into two meridians, cut apart by the poles. One half of the meridians constitutes the east longitudes, which are labelled E or given positive (+) values, and the other half of meridians of longitude are denoted by W and given negative (−) values.

The location of the prime meridian is arbitrary. The idea that it passes through the principal transit instrument, the main telescope, at the Observatory at Greenwich, England, was formally established at the international Meridian Conference in Washington, D.C. There it was decided that there would be a single zero meridian rather than the many used before. Therefore meridian through Greenwich is called the prime meridian. Several other decisions were made at the meeting as well, and among them was the agreement that all longitude would be calculated both east and west from this meridian up to 180° east longitude is positive and west longitude is negative.

The 180° meridian is a unique longitude; like the prime meridian it divides the eastern hemisphere from the western hemisphere. It it also represents the international date line. The calendars west of the line are one day ahead of those east of the line. This division could theoretically occur anywhere on the globe, but it is convenient for it to be 180° from Greenwich in a part of the world mostly covered by ocean.

3.7 Coordinate Transformations

From the preceding section it is clear that there are a number of coordinate systems developed which are in use for specific purposes in GIS. The description of same object in different coordinate systems is different. Often the spatial data captured in one coordinate system need to be transformed to another coordinate system for convenience of computation, visualization, analysis and measurements. The conversion of coordinates transformation from one system to another is performed through mathematical operations. These mathematical transformations of coordinate systems are important cartographic features in GIS systems. In this section mathematics governing different coordinate transformations are discussed.

Therefore for 'n' coordinate systems theoretically there are n(n-1) forward coordinate transformations possible. To consider both the forward and reverse transformations there will be 2n(n-1) transformations possible. Since there are many coordinate systems in use it will be difficult to accomodate all of them in this chapter. Therefore it is prudent to discuss the important transformation methods which are relevent for representing Earth models after classifying them into 2D and 3D categories.

3.7.1 2D Coordinate Transformations

Let the 2D Cartesian coordinate of a point be given by (x, y) and its polar equivalent are given by (r, θ). The frequently used 2D coordinate transformations in GIS are:

1. Cartesian coordinate to polar coordinate transformation
2. Cartesian coordinate to log-polar transformation
3. Cartesian coordinates from bipolar coordinates

The equations governing the computations of Cartesian 2D coordinates from their equivalent polar coordinates are given by

$$x = r cos\theta \tag{3.2}$$

$$y = r sin\theta \tag{3.3}$$

$$\frac{\partial(x,y)}{\partial(r,\theta)} = \begin{pmatrix} cos\theta & -rsin\theta \\ sin\theta & rcos\theta \end{pmatrix} \tag{3.4}$$

$$det\frac{\partial(x,y)}{\partial(r,\theta)} = r \tag{3.5}$$

The reverse transformation of polar coordinates from Cartesian coordinates is given by

$$r = \sqrt{x^2 + y^2} \tag{3.6}$$

$$\theta' = arctan\frac{y}{x} \tag{3.7}$$

The correct value of the angle of the coordinate can be computed from θ' using the logic if ($\theta' < 90°$) then $\theta = \theta'$
else if ($\theta' > 90°$) and ($\theta' < 180°$) then $\theta = \pi - \theta'$
else if ($\theta' > 180°$) and ($\theta' < 270°$) then $\theta = \pi + \theta'$
else if ($\theta' > 270°$) and ($\theta' < 360°$) then $\theta = 2\pi - \theta'$

Also one can use the following pair of equations to compute (r, θ)

$$r = \sqrt{x^2 + y^2} \tag{3.8}$$

$$\theta = 2arctan\frac{y}{x + r} \tag{3.9}$$

To compute the log-polar coordinates (ρ, θ) from Cartesian coordinates (x, y)

$$\rho = log \sqrt{x^2 + y^2} \tag{3.10}$$

$$\theta = arctan\frac{y}{x} \tag{3.11}$$

To compute Cartesian coordinates (x, y) from log-polar coordinates (ρ, θ)

$$x = e^\rho cos\theta \tag{3.12}$$

$$y = e^\rho sin\theta \tag{3.13}$$

3.7.2   3D Coordinate Transformations

Let the Cartesian coordinate in standard three dimension form be represented by (x, y, z) and its equivalent spherical coordinate in the standard form be represented by (ρ, θ, Φ).

Then the frequently used 3D coordinate transformations in GIS are

1. Spherical to 3D-rectangular coordinate transformation
2. Cylindrical to 3D-rectangular coordinate transformation
3. 3D Cartesian coordinate to 3D spherical coordinate transformation

The equations governing the 3D coordinate transformations from spherical coordinates to Cartesian coordinates are given by

$$x = \rho sin\Phi cos\theta \tag{3.14}$$

$$y = \rho sin\Phi sin\theta \tag{3.15}$$

$$z = \rho cos\Phi \tag{3.16}$$

The Jaccobian matrix is given by

$$\frac{\partial(x, y, z)}{\partial(\rho, \Phi, \theta)} = \begin{pmatrix} sin\Phi cos\theta & \rho cos\Phi cos\theta & -\rho sin\Phi sin\theta \\ sin\Phi sin\theta & \rho cos\Phi sin\theta & \rho sin\Phi cos\theta \\ cos\Phi & -\rho sin\Phi & 0 \end{pmatrix} \tag{3.17}$$

The transformation equations to compute the 3D Cartesian coordinates from a general cylindrical coordinates given by (r, θ, h) are given by

$$x = rcos\theta \tag{3.18}$$

$$y = rsin\theta \tag{3.19}$$

$$z = h \tag{3.20}$$

The corresponding Jaccobian matrix is given by

$$\frac{\partial(x, y, z)}{\partial(r, \theta, h)} = \begin{pmatrix} cos\theta & -rsin\theta & 0 \\ sin\theta & rcos\theta & 0 \\ 0 & 0 & 1 \end{pmatrix} \tag{3.21}$$

$$\tag{3.22}$$

To compute the 3D spherical coordinate from the 3D Cartesian coordinate the transformation equations are given by

$$\rho = \sqrt{x^2 + y^2 + z^2} \tag{3.23}$$

$$\Phi = arctan(\frac{y}{x}) \tag{3.24}$$

$$\Phi = arccos(\frac{x}{\sqrt{x^2 + y^2}}) \tag{3.25}$$

$$\Phi = arcsin(\frac{y}{\sqrt{x^2 + y^2}}) \tag{3.26}$$

$$\theta = arctan(\frac{\sqrt{x^2 + y^2}}{z}) \tag{3.27}$$

To compute the spherical coordinates from cylindrical coordinates the transformation equations are given by

$$\rho = \sqrt{r^2 + h^2} \tag{3.28}$$

$$\Phi = \Phi \tag{3.29}$$

$$\theta = arctan(\frac{r}{h}) \tag{3.30}$$

To compute the cylindrical coordinates from Cartesian coordinates the transformation equation are given by

$$r = \sqrt{x^2 + y^2} \tag{3.31}$$

$$\theta = arctan(\frac{y}{x}) + \pi(-x)sgny \tag{3.32}$$

$$h = z \tag{3.33}$$

3.8 Datum Transformation

As assumed by most of us, Earth is not a rigid body. It is a heterogeneous mixture of solid, semi-solid, liquid and gas constantly revolving and spinning around its axis. In addition, a number of secular forces are acting on Earth due to gravitational interactions and magnetic fields.

This dynamic system undergoes many perturbing forces such as tremors, earthquakes, volcanic eruptions, tsunamis and other external as well as internal forces. These forces cause a random change in the structure of the Earth resulting in immeasurable changes. These changes can result in a shifting of the mass balance of the Earth causing a shifting of its center of mass of Earth leading to change in the axis of rotation etc. These changes manifest in the form of physical change in the shape, size, slope, aspect and shift of location of major Earth objects. This phenomenon is depicted in Figure 3.7.

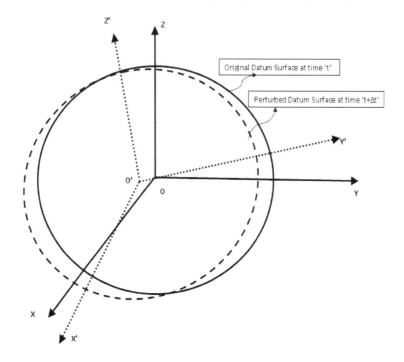

FIGURE 3.7
Transformation of the datum surface

Therefore the modelled datum surface , although appearing constant, keeps changing and varying. This can be thought of as a transformation of the datum surface defined at an epoch of time 't' by (Origin O (x,y,z), semi-major axis (a), semi-minor axis (b), its orientation (α, θ, ϕ)) to a new datum surface $(o', x', y', z' \alpha', \theta', \phi', a', b')$. These changes can be modelled as a conglomeration of rotation, translation and scaling of the ECEF coordinate system which is attached to the original datum surface. This leads to a change in the position of center of mass of Earth as well as in the position of each point on the datum surface.

The change in the datum results in a change in the location measurement and other measurable parameters of the map. To accurately compute the position of the center of mass of Earth as well as the position of special objects with respect to ECEF coordinates, transformations of the datum parameter are necessary. This transformation has been proposed, modeled and experimented by Mikhail Molodensky of Russia and later by Helmert.

3.8.1 Helmert Transformation

Due to the secular forces acting on it, the Earth basically undergoes three types of motion:

- Translational motion
- Rotational motion
- Rotational motion

The transformation due to the translational motion can be easily derived and is given by :

$$X_s = X_r + TX \tag{3.34}$$

$$Y_s = Y_r + TY \tag{3.35}$$

$$Z_s = Z_r + TZ \tag{3.36}$$

Given a sufficient number of points where coordinates are known in both reference systems, the datum shifts TX, TY and TZ can be determined. Else if we can know the shifts (i.e. TX, TY and TZ) we can easily determine the new transformed coordinates. So we can easily apply this transformation for the translational distortion of the Earth.

The above datum transformation model assumes that the axes of the two systems are parallel, the systems have the same scale, and the geodetic network has been consistently computed. Reality is that none of these assumptions occurs, and thus TX, TY and TZ can vary from point to point. A more general transformation involves seven parameters:

1. A change in scale factor
2. The rotation of the axes between two systems.(R_X, R_Y, R_Z)
3. The three translation factor (TX, TY, TZ)

This transformation procedure is known as the Helmert Transformation and it can be written as:

$$
\begin{bmatrix} X \\ Y \\ Z \end{bmatrix} = \begin{bmatrix} TX \\ TY \\ TZ \end{bmatrix} + (1 + \Delta S) \begin{bmatrix} X \\ Y \\ Z \end{bmatrix} + \begin{bmatrix} 0 & R_Z & R_Y \\ -R_Z & 0 & R_X \\ R_Y & -R_X & 0 \end{bmatrix} \tag{3.37}
$$

The parameters are given by:

TX(t) = 0.9910m

TY(t) = -1.9072m

TZ(t) = -0.5129m

$R_X(t) = [125033+258(t-1997.0)](10^{-12})$ radians

$R_Y(t) = [46785 - 3599(\text{t-1997.0})](10^{-12})$ radians

$R_Z(t) = [56529 - 153(\text{t-1997.0})](10^{-12})$ radians

$\Delta S(t) = 0.0$

3.8.2   Molodenskey Transformation

Mikhail Molodensky of Russia has proposed a datum transformation to transform the existing spatial data from previous datum to the modified datum, which is well known as Molodensky Transformation. The equation is given by:

$$f_{WGS84} = f_{Local} + Df \qquad (3.38)$$

$$l_{WGS84} = l_{Local} + Dl \qquad (3.39)$$

$$h_{WGS84} = h_{Local} + Dh \qquad (3.40)$$

Where Df, Dl and Dh are provided by the standard Molodensky Transformation formulas of:

$$Df'' = \{-DX \sin f \cos l - DY \sin f \sin l + DZ \cos f + Da \frac{(R_N + e^2 \sin f \cos f)}{a} +$$
$$Df[R_M(\frac{a}{b} + R_N \frac{b}{a})] \sin f \cos f\}[(R_M + h) \sin 1'']$$

$$Dl'' = [-DX \sin l + DY \cos l] * [(R_N + h) \cos f \sin 1'']^{-1}$$

$$Dh = DX \cos f \cos l + DY \cos f \sin l + DZ \sin f - Da(\frac{a}{R_N}) + Df(\frac{b}{a})$$

where f,l,h are geodetic coordinates of the points on the local datum.

3.9   Usage of Coordinate Systems

From these discussions it can be safely concluded that there are inumerable coordinate systems theorized, modeled and put into practice for proof of many scientific phenomena. The prominent or widely used coordinate systems which are useful in geodesy, geography and GIS are discussed in earlier session. The first and foremost usage of coordinate systems are for the understanding of geometric concepts in mathematics. Therefore, 2D and 3D geometry and related concepts is the basis of understanding of location, direction, distance, area, volume etc. in the Cartesian plane or 3D space. These concepts are further used in physics. The prominent applications of coordinate systems in geodesy are in the form of ECEF, ITRF, UTM and UPS grids. To locate and track military objects in the battle field separate military grid coordinate systems are devised where some of its parameters such as its origin, extend etc. are kept secret for operational secrecy. The celestial coordinate systems and universal coordinate systems are used for tracking of celestial objects viz. planets, stars, comets and satellites.

3.10 Summary

In this chapter the important concept of reference frame or coordinate reference system is discussed. The mathematical definition of a reference frame is given and how the real world spatial data are referenced through a coordinate system by attaching the mathematical model datum to a real world is discussed. Because of the requirement of referring the real world objects, various coordinate system are being developed. The special coordinate systems e.g. rectangular, spherical, cylindrical, polar, log-polar coordinate system are discussed. Further the coordinate system used for referring the Earth such as ECEF, ITRF, UTM, UPS, military grid, celestial coordinate system are discussed. This chapter delves into different transformation of spatial coordinates and their formula. Similarly the datum transformation which is necessary for accurate location of spatial objects on a datum surface is discussed.

4

Basics of Map Projection

This chapter focuses on the key concept of map projection. The sequence of mathematical operations which leads to preparation of maps is explained through a sequence diagram. The important map transformations or map projections and their characteristics are explained. Some of the important applications of map projections are tabulated along with their properties. The classification of map projections from different mathematical perspectives and cartographic aspects is given. This chapter deals with the basic concepts of map projection and answers generic queries such as 'What is map projection?', 'Why is map projection required?', 'Why there are so many projections devised?', 'Which map projections are suitable for a particular application?' etc. How map projections are designed and developed for different regions of the Earth's surface is explained. Classification of different map projections using different development surfaces, perspective positions and positions of tangent surface is explained. The categorization of map projections under different categories such as development surface, usage and characteristics is discussed. The mathematical formulae for forward and reverse map projections are derived ab initio. The chapter ends with a discussion on how to choose a map projection for a specific purpose and for a specific region of the world. All the concepts introduced here are supplemented by illustrations and key notes.

4.1   What Is Map Projection? Why Is It Necessary?

A map is a two-dimensional piece-wise representation of the Earth's crust on a paper surface (Snyder 1989; 1994). The shape of the Earth cannot be equated to any conventional geometric shapes i.e. it cannot be represented or modeled to any standard geometric shape like a sphere, ellipse etc. Depicting Earth's surface on a two-dimensional paper surface without distortion is akin to pasting an orange peel on the table surface so as to get a uniform and continuous strip on the surface without tearing and stretching the peel, in other words, a near impossible job. But cartographers and scientists have devised a number of mathematical formulae to accomplish the job. For the map projections the Earth's surface is modeled as an ellipsoid, spheroid or geoid for different purposes. Actually, the Earth is more nearly an oblate spheroid - an ellipse rotated about its minor axis. Then an appropriate mathematical formula is designed so as to represent the modeled Earth's surface on a flat paper surface. These mathematical formulae, which project or translate the Earth's surface to paper surface with minimum distortion, are known as map projections. Earth is modeled as a globe for numerous cartographic reasons, but it is not possible to make a globe on a very large scale. If anyone wants to make a globe on a scale of one inch to a mile, the radius of the globe will have to be 330 ft. It would be difficult to make and handle such a globe and uncomfortable to carry it to the field for reference. A small globe is useless for referring to a small country or landscape because it distorts the smaller land surfaces and depicts the land surfaces inappropriately. So for practical purposes a globe is least useful or helpful in the field. Moreover it is neither easy to compare in detail different regions of the Earth over the globe, nor is it convenient to measure distances over it. Hence maps were devised to overcome such difficulties. A map is a two-dimensional representation of a globe drawn on paper map which is convenient to fold and carry in the field and easy to compare and locate different parts of the Earth. Locating a known feature, guiding and navigating from one position to another and comparing two different regions over a map are convenient and easy. Transforming a three-dimensional globe to a two-dimensional paper map is accomplished using map projection. Topographical maps of different scales, atlases and wall maps are prepared using map projections. Thus map projection plays a crucial role in preparation of different types of maps with different scales, coordinate systems and themes.

4.2   Mathematical Definition of Map Projection

A map projection is defined as a mathematical function or formula which projects any point (ϕ, λ) on the spherical surface of Earth to the two-dimensional point (x, y) on a plane surface. The forward map projection is given by

$$(x, y) = f(\phi, \lambda) \tag{4.1}$$

Often the geographic data obtained in Cartesian coordinate needs to be transformed to spherical coordinates. This necessitates the reverse process which is known as inverse map projection. The inverse map projection is given by

$$(\phi, \lambda) = f^{-1}(x, y) \tag{4.2}$$

Thus, the mathematical function, which realizes the map projection, essentially projects the 3D spatial features onto 2D map surfaces. Invaluable re-

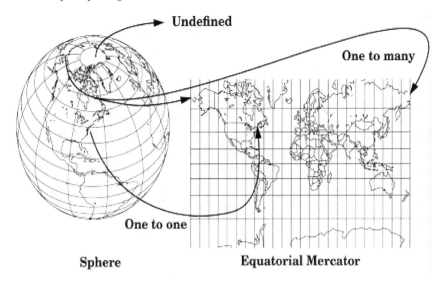

FIGURE 4.1
Map projection, the mapping of Earth coordinates to map coordinates

sources for the mathematical formulations of different map projections are Snyder [52], [54], and Fenna [15]. From Figure 4.1 it can be concluded that map projection is a systematic mapping of points on the globe to a 2D rectangular surface.

Another definition of map projection is a mathematical equation or series of equations, which takes a three-dimensional location on the Earth and provides corresponding two-dimensional coordinates to be plotted on a paper or computer screen.

Map projections are treated mathematically as the transformation of geodetic coordinates (ϕ, λ) into rectangular grid coordinates often called easting and northing. This transformation is expressed as a series of equations and implemented as a computer algorithm.

4.3 Process Flow of Map Projection

Figure 4.2 depicts the flow of cartographic process for map projection. The real Earth is modelled with a suitable mathematical surface known as the datum surface or the geodatic datum. The datum surface is scaled down. The scaled down datum surface is warped with a suitable geometric surface.

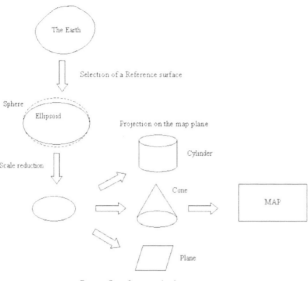

Process flow of map projection

FIGURE 4.2
Process flow of map projection

The geometric surface is unwarped to prepare the map using suitable map projection techniques.

4.4   Azimuthal Map Projection

See Figure 4.3 for azimuthal projection.

- 'O' is the center of the the Earth i.e. scale reduced of the Earth with a spherical datum.

- 'R' is the mean radius of the Earth.

- 'V' is the view point or eye point of the observer.

- 'F' is the point representing the pole of the Earth which concides with the point on the map container.

- α is the view angle i.e. the angle subtended by the eye looking at the point 'P'.

- $P(\Phi, \lambda)$ is a generic point on the surface of the Earth.

- β is the angle subtended by chord PF at the center of the Earth.

- $P^{'}(x, y)$ is the azimuthal projection of point 'P' on the map surface.

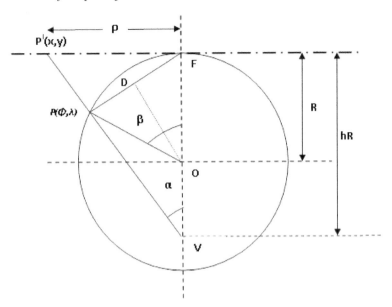

FIGURE 4.3
Schematic of azimuthal map projection

In $\Delta P'VF$ the radial distance i.e. $P'F = \rho$ is given by the equation

$$P'F = \rho = hR\tan\alpha \tag{4.3}$$

In Δ POF PD = DF as OD is the perpendicular bisector of the chord PF
$\angle PDG = \angle FDG = \frac{\pi}{2}$
Using the SAS (Side-Angle-Side) rule for congruency of triangles
In $\triangle ODP \cong \triangle ODF$
$\Rightarrow \angle POD = \angle FOD = \frac{\beta}{2}$
Now PD = DF = R $\sin\frac{\beta}{2}$

$$PF = 2R\sin\frac{\beta}{2} \tag{4.4}$$

Now $\angle$ DFO = $\angle$ DPO = $\frac{\pi}{2} - \frac{\beta}{2}$
Now in ΔVPF

$$\angle VPF = \Pi - (\alpha + \frac{\pi}{2} - \frac{\beta}{2}) \tag{4.5}$$

$$\Rightarrow \angle VPF = \frac{\pi}{2} + \frac{\beta}{2} + \alpha \tag{4.6}$$

Now applying sine rule in ΔVPF

$$\frac{VF}{\sin \angle VPF} = \frac{PF}{\sin \angle PVF} \tag{4.7}$$

Now substituting the values from equation 4.6

$$\frac{hR}{\sin\frac{\pi}{2} + \frac{\beta}{2} + \alpha} = \frac{PF}{\sin\angle PVF} \tag{4.8}$$

$$\Rightarrow h\sin\alpha = 2\sin\frac{\beta}{2}\cos(\frac{\beta}{2} - \alpha) \tag{4.9}$$

$$\Rightarrow h\sin\alpha = 2\sin\frac{\beta}{2}\cos\frac{\beta}{2}\cos\alpha + 2\sin^2\frac{\beta}{2}\sin\alpha \tag{4.10}$$

$$\Rightarrow \sin(h - 2\sin^2\frac{\beta}{2}) = \cos\alpha\sin\beta \tag{4.11}$$

$$\Rightarrow \tan\alpha = \frac{\sin\beta}{h - 2\sin^2\frac{\beta}{2}} \tag{4.12}$$

Now putting the value of $\tan\alpha$ in the equation one can obtain

$$\rho = \frac{hR\sin\beta}{h - 2\sin^2\frac{\beta}{2}} = \frac{hR\sin\beta}{h - 1 + \cos\beta} \tag{4.13}$$

$$\Rightarrow \rho = \frac{hR\cos\Phi}{h - 1 + \sin\Phi} \tag{4.14}$$

Now substituting the radial distance in the equation 4.14 one can obtain

$$x = \rho\sin(\lambda - \lambda_x) \tag{4.15}$$

$$\Rightarrow x = (\frac{hR\sin\beta}{h - 1 + \cos\beta})\sin(\lambda - \lambda_x) \tag{4.16}$$

$$y = \rho\cos(\lambda - \lambda_x) \tag{4.17}$$

$$\Rightarrow y = (\frac{hR\sin\beta}{h - 1 + \cos\beta})\cos(\lambda - \lambda_x) \tag{4.18}$$

4.4.1   Special Cases of Azimuthal Projection

Specific instances of the azimuthal projection can be obtained by substituting the values of 'h' one can obtain the following specific projections.

- For $h = 1$ The projection is called gnomonic projection
- For $h = 2$ The projection is called stereographic projection
- For $h = \infty$ The projection is called orthographic projection

For gnomonic projection i.e. h=1

$$x = R\tan\beta\sin(\lambda - \lambda_x) = R\cot\Phi\sin(\lambda - \lambda_x) \tag{4.19}$$

$$y = R\tan\beta\cos(\lambda - \lambda_x) = R\cot\Phi\cos(\lambda - \lambda_x) \tag{4.20}$$

For stereographic projection i.e. h=2

$$x = 2R \tan \frac{\beta}{2} \sin(\lambda - \lambda_x) = 2R \cot \Phi \sin(\lambda - \lambda_x) \tag{4.21}$$

$$y = 2R \tan \frac{\beta}{2} \cos(\lambda - \lambda_x) = 2R \cot \Phi \cos(\lambda - \lambda_x) \tag{4.22}$$

For orthographic projection i.e. h=∞

$$x = \lim_{h \to \infty} \left(\frac{hR \sin \beta}{h - 1 + \cos \beta} \right) \sin(\lambda - \lambda_x) \tag{4.23}$$

$$= \lim_{h \to \infty} \left(\frac{hR \cos \Phi}{h - 1 + \sin \Phi} \right) \sin(\lambda - \lambda_x) \tag{4.24}$$

$$= \lim_{h \to \infty} \left(\frac{R \cos \Phi}{1 - \frac{1}{h} + \frac{\sin \Phi}{h}} \right) \sin(\lambda - \lambda_x) \tag{4.25}$$

$$= R \cos \Phi \sin(\lambda - \lambda_x) \tag{4.26}$$

$$similarly : y = R \cos \Phi \cos(\lambda - \lambda_x) \tag{4.27}$$

4.4.2 Inverse Azimuthal Projection

Inverse azimuthal projection transforms the Cartesian coordinates of a point on a map to the spherical coordinates of the datum surface of the Earth.

$$x^2 + y^2 = \left(\frac{hR \cos \Phi}{h - 1 + \sin \Phi} \right)^2 \tag{4.28}$$

For gnomonic projection i.e. h=1

$$\sqrt{x^2 + y^2} = \frac{R \cos \Phi}{\sin \Phi} \tag{4.29}$$

$$\Rightarrow \Phi = \cot^{-1} \left(\frac{\sqrt{x^2 + y^2}}{R} \right) \tag{4.30}$$

For stereographic projection i.e. h=2

$$\sqrt{x^2 + y^2} = \frac{2R \cos \Phi}{1 + \sin \Phi} \tag{4.31}$$

$$\Rightarrow \frac{\sqrt{x^2 + y^2}}{2R} = \frac{\frac{1 - \tan^2 \frac{\Phi}{2}}{1 + \tan^2 \frac{\Phi}{2}}}{1 + \frac{2 \tan \frac{\Phi}{2}}{1 + \tan^2 \frac{\Phi}{2}}} \tag{4.32}$$

$$= \frac{1 - \tan^2 \frac{\Phi}{2}}{1 + \tan^2 \frac{\Phi}{2} + 2 \tan \frac{\Phi}{2}} \tag{4.33}$$

$$= \frac{\frac{1 - \tan \frac{\Phi}{2}}{1 + \tan \frac{\Phi}{2}}}{(1 + \tan \frac{\Phi}{2})^2} \tag{4.34}$$

$$= \frac{1 - \tan\frac{\Phi}{2}}{1 + \tan\frac{\Phi}{2}} \tag{4.35}$$

$$= \tan(\frac{\Pi}{4} - \frac{\Phi}{2}) \tag{4.36}$$

$$\Rightarrow \Phi = \frac{\Pi}{2} - 2\tan^{-1}(\frac{\sqrt{x^2 + y^2}}{2R}) \tag{4.37}$$

For orthographic projection i.e. $h = \infty$

$$x^2 = R^2 cos^2\Phi \sin^2(\lambda - \lambda_x) \tag{4.38}$$
$$y^2 = R^2 cos^2\Phi \cos^2(\lambda - \lambda_x) \tag{4.39}$$

$$\Rightarrow x^2 + y^2 = R^2 \cos^2\Phi \tag{4.40}$$

$$\Rightarrow \Phi = \cos^{-1}(\frac{x^2 + y^2}{R^2}) \tag{4.41}$$

Using the sub equations from equation 4.39 and dividing them one can obtain

$\frac{x}{y} = \tan(\lambda - \lambda_x)$
$\Rightarrow \lambda - \lambda_x = \tan^{-1}\frac{x}{y}$

$$\lambda = \lambda_x + \tan^{-1}\frac{x}{y} \tag{4.42}$$

4.5   Cylindrical Map Projection

The diagram and equations developed for projecting onto the azimuthal plane are applied similarly to the cylindrical situation (see Figure 4.4).

If the double dotted lines are seen as Earth's axis and dashed line as the equator then

$$y = \rho = hR\tan\alpha \tag{4.43}$$

Asuming the central meridian lies along Y axis with north at the top of the map, the generic equations for the cylindrical mode at simple aspects are

$$x = R\lambda \tag{4.44}$$
$$y = hR\tan\alpha \tag{4.45}$$

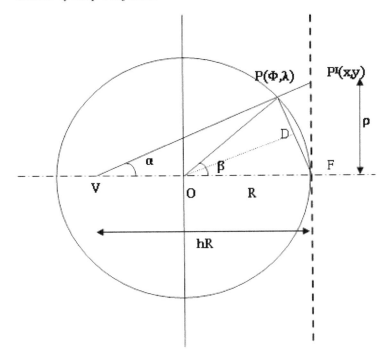

FIGURE 4.4
Schematic of cylindrical map projection

From the previous calculations of azimuthal projection we have
$\tan \alpha = \frac{\sin \beta}{h - 1 + \cos \beta}$
For the cylindrical coordinate system $\beta = \Phi$
$\Rightarrow \tan \alpha = \frac{\sin \Phi}{h - 1 + \cos \phi}$
Substituting the value of $\tan \alpha$ we get

$$x = R\lambda \qquad (4.46)$$

$$y = \frac{hR \sin \Phi}{h - 1 + \cos \Phi} \qquad (4.47)$$

4.5.1 Special Cases of Cylindrical Projection

Like the case of azimuthal map projection we can derive the special cases for the cylindrical projection i.e.

- Gnomonic projection
- Stereographic projection
- Orthographic projection

4.5.1.1 Gnomonic Projection

In gnomonic projection the observer is at the center of the Earth i.e. h = 0.

$$x = R\lambda \tag{4.48}$$
$$y = R\tan\Phi \tag{4.49}$$

4.5.1.2 Stereographic Projection

In stereographic projection the observer is placed at opposite of the diameter of the Earth i.e the value of h = 2.

$$x = R\lambda \tag{4.50}$$
$$y = 2R\tan\frac{\Phi}{2} \tag{4.51}$$

4.5.1.3 Orthographic Projection

In orthographic projection the observer stands at infinite distance from the surface of the Earth. Therefore the value of $h = \infty$.

$$x = R\lambda \tag{4.52}$$

Now
$$y = \lim_{h\to\infty}\left(\frac{hR\sin\Phi}{h-1+\cos\Phi}\right)$$
$$y = \lim_{h\to\infty}\left(\frac{R\sin\Phi}{1-\frac{1}{h}+\frac{\cos\Phi}{h}}\right)$$

$$\Rightarrow y = R\sin\Phi \tag{4.53}$$

4.5.2 Inverse Transformation

Here we will make the inverse transformation from cylindrical coordinate system to the Cartesian coordinate sysytem.

$$\lambda = \frac{x}{R}$$
$$y = \frac{hR\sin\Phi}{h-1+\cos\Phi}$$

For gnomonic projection i.e. h=1
$$y = \frac{R\sin\Phi}{\cos\Phi}$$

$$\Rightarrow \Phi = \tan^{-1}\frac{y}{R} \tag{4.54}$$

For stereographic projection i.e. h=2
$$y = \frac{2R\sin\Phi}{1+\cos\Phi}$$

$$\Rightarrow \frac{y}{2R} = \frac{\frac{4\tan\frac{\Phi}{2}}{1+\tan^2\frac{\Phi}{2}}}{1+\frac{1-\tan^2\frac{\Phi}{2}}{1+\tan^2\frac{\Phi}{2}}}$$

$$= 2\tan\frac{\Phi}{2}$$

$$\Rightarrow \Phi = 2\tan^{-1}\frac{y}{4R} \tag{4.55}$$

For orthographic projection we have $h = \infty$
$y = R\sin\Phi$

$$\Rightarrow \Phi = \sin^{-1}\frac{y}{R} \tag{4.56}$$

4.6 Conical Map Projection

Figure 4.5 shows that conic map projection has a lot of similarity to the diagram and equations for cylindrical map projection. Therefore conical map projection can be developed with slight modification and difference. The taper of the cone is characterised by the ratio of its cross sectional radius to the length of the slope from its apex. This ratio is constant for any distance from the apex of a given cone and is called constant of the cone c given by the equation

$$c = sin\alpha \tag{4.57}$$

where α is the semi-apex angle of the cone. Also angle AGF forms the complement for both α and Φ_0, Therefore $\alpha = \Phi_0$

$$c = sin|\Phi_0| \tag{4.58}$$

If the distance along slope from the apex of the cone is ρ, cross-section radial length is given by

$$radial - length = c\rho \tag{4.59}$$

In the triangle GAF

$$\rho_0 = Rcot\alpha \tag{4.60}$$

To develop a cone into a map requires cutting it lengthwise so as to flatten the surface (Figure 4.6). Assume s is the central meridian and cut down a ray on the opposite side of the cone. The resulting map has the shape of a sector of disk with A as the center. So ray carrying mapped points of any one meridian becomes radial line of sector and angle θ between any two ray becomes c times the angle at poles between their respective meridians.

$$\theta = c\lambda \tag{4.61}$$
$$X = \rho sin\theta = \rho sin(c\lambda) \tag{4.62}$$
$$Y = \rho_0 - \rho_1 cos\theta = Rcot\phi_0 - \rho_0 cos(c\lambda) \tag{4.63}$$

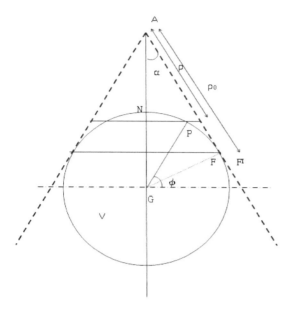

FIGURE 4.5
Schematic of conical map projection

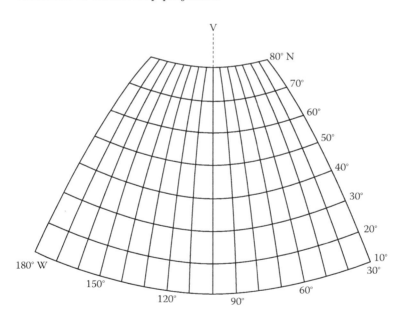

FIGURE 4.6
Flattened cone after cutting along a central meridian

Now by analyzing the figure one can derive

$$PF = \frac{hRsin(\phi - \phi_0)}{h - 1 + cos(\phi - \phi_0)} \tag{4.64}$$

$$\rho = \rho_0 - PF = RCot\phi_0 - \frac{hRsin(\phi - \phi_0)}{h - 1 + cos(\phi - \phi_0)} \tag{4.65}$$

$$X = \rho sin(c\lambda) = R(cot\phi_0 - \frac{hRsin(\phi - \phi_0)}{h - 1 + cos(\phi - \phi_0)})sin(c\lambda) \tag{4.66}$$

$$Y = Rcot\phi_0 - R(cot\phi_0 - \frac{hsin(\phi - \phi_0)}{h - 1 + cos(\phi - \phi_0)})cosc\lambda \tag{4.67}$$

The gnomonic, stereographic and orthographic versions of the conical projection can be obtained from the above equations by substituting the value of $h = 1$, $h = 2$ and $h = \infty$ respectively. By rearranging the above forward conic projection equations one can obtain

$$X = \rho sinc\lambda = R(cot\phi_0 - \frac{hsin(\phi - \phi_0)}{h - 1 + cos(\phi - \phi_0)})sinc\lambda \tag{4.68}$$

$$Rcot\phi_0 - Y = R(cot\phi_0 - \frac{hsin(\phi - \phi_0)}{h - 1 + cos(\phi - \phi_0)})cosc\lambda \tag{4.69}$$

The inverse transformation of the conic transformation can be obtained by squaring and adding the equations

$$X^2 + (Rcot\phi_0 - y)^2 = R^2(cot\phi_0 - \frac{hsin(\phi - \phi_0)}{h - 1 + cos(\phi - \phi_0)})^2 \tag{4.70}$$

$$\frac{Rhsin(\phi - \phi_0)}{h - 1 + cos(\phi - \phi_0)} = Rcot\phi_0 - \sqrt{(x^2 + (Rcot\phi_0 - Y)^2)} \tag{4.71}$$

By substituting $h = 1$ in the above equation one can obtain

$$Rtan(\phi - \phi_0) = Rcot\phi_0 - \sqrt{(x^2 + (Rcot\phi_0 - Y)^2)} \tag{4.72}$$

$$\phi = \phi_0 + tan^{-1}(\frac{Rcot\phi_0 - \sqrt{x^2 + (Rcot\phi_0 - Y)^2}}{R}) \tag{4.73}$$

For stereographic projection, by substituting $h = 2$ in the above equation one can obtain

$$\frac{2Rsin(\phi - \phi_0)}{1 + cos(\phi - \phi_0)} = Rcot\phi_0 - \sqrt{x^2 + (Rcot\phi_0 - Y)^2} \tag{4.74}$$

$$2Rtan\frac{\phi - \phi_0}{2} = Rcot\phi_0 - \sqrt{x^2 + (Rcot\phi_0 - Y)^2} \tag{4.75}$$

$$\phi = \phi_0 + 2tan^{-1}(\frac{Rcot\phi_0 - \sqrt{x^2 + (Rcot\phi_0 - Y)^2}}{2R}) \tag{4.76}$$

For orthopgraphic projection, by substituting $h = \infty$ in the above equation one can obtain

$$R sin(\phi - \phi_0) = R cot\phi_0 - \sqrt{x^2 + (R cot\phi_0 - Y)^2} \qquad (4.77)$$

$$\phi = \phi_0 + sin^{-1}\left(\frac{R cot\phi_0 - \sqrt{x^2 + (R cot\phi_0 - Y)^2}}{R}\right) \qquad (4.78)$$

The value of meridian λ can be obtained from

$$Tanc\lambda = \frac{X}{R cot\phi_0 - Y} \qquad (4.79)$$

$$\lambda = \frac{1}{c} tan^{-1} \frac{X}{R cot\phi_0 - Y} \qquad (4.80)$$

4.7 Classification of Map Projections

After studying the standard map projection methods and their mathematical derivations one can state that map projections are functions of (ϕ, λ) and datum parameters. A datum surface is a mathematical model of Earth. Also it is clear that no one datum parameters can exactly model the shape of the Earth accurately. Because infinite numbers of datum are possible, therefore theoretically infinite numbers of map projections are possible for the same value of (ϕ, λ) but different datum parameters.

Therefore out of the theoretically infinite possible map projections it is necessary to find the map projections which are highly accurate and practically useful for the purpose. This challenging job has been simplified by listing finitely many datum which are practically useful and give a good approximation of the shape of Earth for a particular region. The list of useful datum is given in Table 13.1 of the Appendix. WGS-84 is one of the best fit datum globally accepted and is being used by satellite navigation systems worldwide.

One way of finding a useful map projection out of the many theoretical possibilities is to classify them into few categories such that each category of the projections exhibits some common property. The classification is possible because many of the map projections have common cartographic, geometric and physical properties. The map projection classification method based on the criteria and their characteristics is discussed in this section.

Map projection varies with size and location of different areas on the surface of Earth. While cylindrical projection is appropriate for the equatorial region, conical projection gives the best fit for tropical to polar regions and azimuthal projection is best for polar region surrounding the pole. Not only that, map projection varies with respect to the purpose for which the map is to be used. While transferring the datum surface to the map surface there are certain things that should be kept in view.

Various important criteria that forms the basis of classification of map projections are listed in Table 4.7.

Criteria of Map Transformation	Classes of Map Projection
Depending on preservation of cartographic measurements	Distance preserving, direction preserving, area preserving
Depending upon the location of the viewer or the light source with respect to the projection surface	Gnomonic, stereographic or orthographic
Depending on the geometry of the projection surface	Cylindrical, conical, planner
Depending on the placement of the projection surface with respect to the datum surface	Tangent or secant
Depending on the orientation of the projection surface with respect to the datum surface	Orthogonal, oblique, transverse

TABLE 4.1
Criteria of Projecting Earth Surface and Classes of Map Projections

4.7.1 Classification Based on the Cartographic Quantity Preserved

Map projection is the mathematical transformation of the 3D surface of Earth to a 2D planar map. It is clear from this mathematical definition that map projections are always erroneous transformations. The fundamental cartographic quantities that can be measured on a map are the area, shape, distance or direction. A single map projection can not give correct measurements of these three quantities simultaneously on a map. These are mutually exclusive cartographic measurements in a projected map. A map projection can preserve one of the quantities accurately at the expense of the other quantities. Therefore map projections can be classified into three classes on the basis of the cartographic quantity they preserve. The classes of map projection depending on the cartographic quantity they preserve and their likely area of applications are listed below.

1. **Equal area or homolographic projection:** This projection preserves the area measurement on the map. In this projection system
   the graticules are prepared in such a way that every quadrilateral
   on it may appear proportionally equal in area to the corresponding
   spherical quadrilateral. These map projections are useful for applications involving estimation of land use, land cover measurement,
   land partitioning, city and town planning, area-wise resource allo-
   cation, cadastral applications etc.

2. **Orthomorphic projection:** These types of projections preserve
   the shape of the mapping surface. They are also known as conformal projections. It is relatively difficult to preserve the shape of
   large mapped area. But the shape of small areas are preserved. So
   in order to increase the quality of orthomorphism certain modifica-
   tions are carried out in the projection. These projections are useful
   for preparation of small scale maps where the overall shape of the
   land mass has to be preserved such as the preparation of political
   maps.

3. **Azimuthal projection:** These types of map projections preserve
   the direction or bearing of one point from the other point of the
   surface. In this type of projection the true bearing or the azimuths
   are preserved. This can be done most efficiently by zenithal projection, in which the datum surface is viewed either from the center (gnomonic projection), or from the anti point of center (stereographic projection) or from infinity (orthographic projection). For
   the map to show all directions correctly, the rectangular quality
   of the spherical quadrilateral as well as the true proportion of its
   length and breadth is maintained. The direction and distance preserving map projection is quite useful for preparation of maps for
   navigational applications. Long distance navigation charts of land,
   sea and air are preparted using azimuthal projections.

4.7.2   Classification Based on the Position of the Viewer

Map projections can be classified depending on the position of the viewer with
respect to the datum surface. In cartographic science a light source is kept
with respect to the viewer position. The point of projection is derived from
the correspondence established by the ray of light which cuts the transparent
datum surface at a point and falls on the map surface at a particular unique
location. Depending on the position of the light source at the center of the
datum surface, anti podal or at deep space with respect to the projection
plane, the projections are known as gnomonic, stereographic or orthographic
respectively. The three types of map projections are depicted in Figure 4.7.

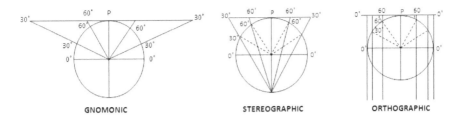

FIGURE 4.7
Map projections based on the position of the viewer

1. **Gnomonic:** In gnomonic projection which is also known as the central projection the observer is placed at the center of the Earth. This projection is most suitable for the point on the polar region to the pole.

2. **Stereographic:** In this type of projection which is also known as perspective transformation the observer can be thought of at any anti point of the center, i.e. at poles. If perfectly calibrated this can give a conformal projection. This projection gives the best fit for the tropical region of the Earth.

3. **Orthographic:** This projection is also known as the deep space projection. The observer observes the datum surface from deep space. The lines of sight fall perpendicular to the mapping surface and are parallel sets of lines from a position far away from the datum surface. This projection is best suited for satellite photography of the Earth surface.

4. **Others:** This includes the generic type of projection, where the observer can be anywhere on the datum surface. This can give promising results if we want to find the projection of a point with respect to any arbitrary point on the datum surface.

4.7.3 Classification Based on Method of Construction

1. **Perspective:** In perspective projection the lights coming from the object (in this context the map surface or the datum surface) converges at a particular points, and the point of convergence is the best position for the eye of the observer. gnomonic projection, stereographic projections etc. the examples of the perspective projection.

2. **Non-perspective:** In non-perspective projection the lights coming from the map surface or the datum surface converges at infinity, i.e. they are parallel to each other. Orthographic projection is an example of non-perspective projection.

4.7.4   Classification Based on Developable Map Surface

One of the important criteria of classifying map projections is based on the
geometry of the projection surface. The geometrical surfaces which are used
in cartography for projection of the datum surface are cylinder, plane or cone.
These are known as map developable surfaces. Therefore depending on the
type of the developable surface map projections can be classified into cylindrical, conical, planar or azimuthal projection. Also depending upon the contact
of the developable surface with that of the datum surface the map projections are further classified into tangent, secant, normal or oblique type. These
classes of map projections are depicted in Figure 4.8 and 4.9.

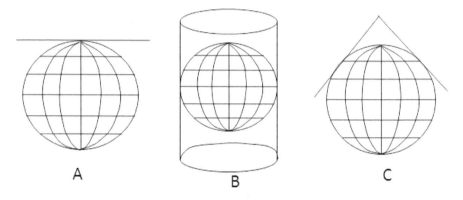

FIGURE 4.8
Geometry of map developable surfaces: (A) planar, (B) cylindrical, (C) conical
placed tangent to the datum surface

1. **Cylindrical:** In cylindrical map projection the development sur-
    face is a cylinder wrapped around the datum surface and is normal
    to the equatorial plane. Cylindrical projections are best suited for
    the equatorial region and are widely used. Mercator projection is a
    cylindrical projection used universally for many applications.

2. **Azimuthal:** In this type of projection the map surface is a plane
    tangent or secant to the datum surface. This type of projection is
    best suited for the polar regions of the Earth.

3. **Conical:** In this class of projections the map surface is assumed to
    be a cone wrapped around the datum surface. The surface of the
    cone can be tangent or secant to the datum surface. Often multiple
    cones are fitted to a region and the cone surface is unfolded to
    develop the map. Lambert Conformal Conic (LCC) projection is an
    example of conical projection. This class of projection is best suited
    for the regions in between the pole and the equator i.e. the tropical
    zone of the Earth's surface.

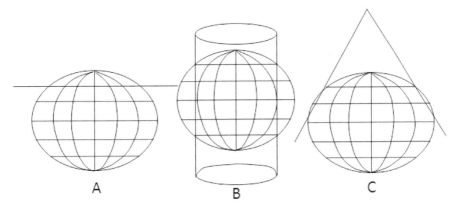

FIGURE 4.9
Geometry of map developable surfaces: (A) planar, (B) cylindrical, (C) conical placed secant to the datum surface

4.7.5 Classification Based on the Point of Contact

Map projections can be classified based on the point of contact of the map surface with that of the datum surface. Based on the point of contact of the tangent surface with the datum surface three classes of map projections are derived viz. polar, equatorial and oblique as depicted in Figure 4.10.

1. **Polar:** In this type of projection the map surface is a tangent to the datum surface and the point of contact is the north pole or

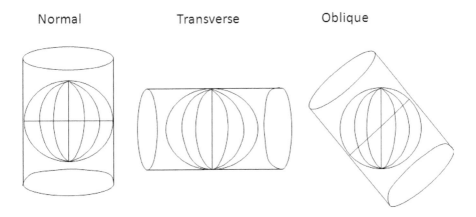

FIGURE 4.10
Geometry of the map projections depending upon the orientation of the map surface with the datum surface: (A) normal, (B) transverse, (C) oblique

the south pole depending upon the requirement. Therefore the map surface is parallel to the equatorial plane.

2. **Equatorial or Normal:** In normal map projection the map surface is a tangent to the datum surface, and is perpendicular to the plane of equator.

3. **Oblique:** In oblique map projection the map surface touches the datum surface at any angle with the equatorial plane [53]. This type of projection is mostly useful to find the projection of a particular area with respect to another point on the datum surface.

4.8   Application of Map Projections

4.8.1   Cylindrical Projections

Mercator projection is a cylindrical map projection having a conformal property. This map projection is used to display accurate compass bearings for navigation at sea. Therefore any straight line drawn on a map projected using the Mercator projection represents a line with each point having constant compass bearing or true direction. A line on a map having constant bearing is known as a 'loxodrome' or 'rhumb' line. Rhumb lines are useful for navigation at sea with only the aid of the map and compass. The sailing direction or course along the great circle changes constantly and at every moment. A great circle line is also known as an 'Orthodrome'. One of the best known and widely used cylindrical projections is the Mercator cylindrical projection which is used for topographical mapping [52].

4.8.1.1   Universal Transverse Mercator (UTM)

Universal Transverse Mercator (UTM) or Transverse Mercator projections are best known examples of normal cylindrical projections. UTM is an equidistant cylindrical projection also known as 'Plate Carree' projection. Plate Carree projection is used for projecting the world in its entirety.

UTM uses a transverse cylinder fitted as secant to the reference surface. This projection was highly recommended for topographic mapping by the United Nations cartographic committee in 1952. The UTM divides the world into 60 narrow longitudinal zones (see Figure 3.6) of 6 degrees each numbered 1 to 60. The narrow zones of 6 degrees and the secant surface to the datum surface make the distortions very small such that the error can be ignored while preparing large scale maps of 1:10,000.

UTM is designed to cover the world excluding the Artic and Antarctic regions. The areas not included in the UTM system i.e. the regions north of

80° north and south of 80° south are mapped using UPS (Universal Polar Stereographic) projection. The figure shows the UTM zones indexing system. Each zone index constitutes a 'number-alpha' index scheme where the number refers to the 8° zone and / or the column designator and the character ('A'-'X') are used as the row designator of the UTM zones. Each UTM zone has a central meridian which has a scale factor of 0.9996.

In order to avoid negative coordinates for portions located west of the central meridian the central meridian has been assigned a (false) easting value of 5000,000 (meters). The equator has been given a northing value of '0' meter. For measuring the positions north of the equator a (false) northing value of 10,000,000 meters has been assigned to the equator for designating positions south of the equator. If a map series crosses more than one UTM zone then the easing value will change suddenly at the zone junction. To overcome this problem a 40 km overlap into the adjacent zone is allowed.

4.8.1.2   Transverse Mercator projection

Transverse Mercator projection is a transverse cylindrical projection of the datum surface. This is also known by the name 'Gauss-Kruger' or 'Gauss-conformal' projection. For small area maps prepared by transverse Mercator projection the angles and slopes are measure correctly as a result of conformality.

Transverse Mercator projection is useful in many countries for local map preparation when high-scale mapping is carried out. Pan-Europe mapping at large scale 1:500, 000 is carried out using transverse Mercator projection.

4.8.1.3   Equidistant Cylindrical Projection

An Equidistant Cylindrical Projection (ECP), also known as cylindrical projection, simple cylindrical projection or 'Plate Carree' has a true scale along all meridians. Google Earth uses ECP for displaying its images. The transverse version of ECP is also known as Cassini projection. Lambert's Cylindrical Equal Area (LCEA) projection represents areas correctly but does not have noticeable sharp distortions towards the polar region.

4.8.1.4   Pseudo-Cylindrical Projection

Pseudo-cylindrical projections are map projections in which the parallels are represented by parallel straight lines and meridians by curves. Pseudo-cylindrical projections are equal area projections. Robinson's projection is an example of a pseudo-cylindrical projection which is neither conformal nor equal-area but the distortions are very low within about 45° of the center and along the equator. Therefore Robinson's projection which is more frequently used for thematic world mapping provides a more realistic view of the world than rectangular maps such as Mercator.

4.8.2   Conic Map Projection

Conic map projection has following four variants which are well known and used by cartographic community for various applications.

1. Lambert's Conformal Conic (LCC)
2. Simple conic projection
3. Albers equal area
4. Polyconic

These projections are very useful for mapping mid latitude regions of Earth's surface and for countries that do not have span extent in latitude.

4.8.2.1   Lambert's Conformal Conic

The properties of LCC projections which make it useful for small scale mapping of mid latitude regions such as India are:

1. LCC is a conformal map projection.
2. Parallels and meridian intersect at right angle.

Though areas in the maps are inaccurate in conformal projections it is widely used for topographic maps of 1:500,0000 scale.

4.8.2.2   Simple Conic Projection

A simple conic is a normal conformal projection with one standard parallel. The scale of the maps is true scale along all meridians. It produces maps which are equidistant along the meridians. In these projections both shape and area are reasonably well preserved. Russia and Europe are better portrayed on conic projection with two standard parallels.

4.8.2.3   Albers Equal Area Projection

Albers equal-area projection uses two standard parallels. It represents areas correctly with reasonable shape distortion between the standard parallels. This projection is best suited for mapping regions predominantly with East-West extent and located land masses in mid latitude regions. Albers equal-area projection is used for mapping the US thematic map and for the preparation of world atlases.

4.8.2.4   Polyconic Projection

Polyconic projections are neither conformal nor equal area. It is projected into ones tangent to each parallel so that meridians are curved not straight lines.

Scale is true only along the central meridian and parallel. The meridional scale increases as we proceed away from the central meridian. The projection is not suitable beyond 30° on each side of the central meridians. The shape gets distorted in polar regions. Therefore it is restricted within 20° of pole. Polyconic projection is useful for large-scale mapping of the US and for preparation of costal charts by US coast geodetic survey.

4.8.3 Azimuthal Projections

Azimuthal or zenithal or planer map projection is a plane tangent (or secant) to the reference surface. All azimuthal projections possess the property of maintaining correct azimuth or true directions from the center of the map. In the polar cases, the meridians radiate out from the pole at correct angular distance. The azimuthal projections assumes the position of the light source or the map viewer into three cardinal points of the datum surface. Depending on the position of the viewer there are three well known types of map projections viz.

1. Gnomonic map projection where the perspective point is at the center of the Earth.

2. Stereographic map projection where the perspective center is at the opposite pole to the point of tangency of the map surface.

3. Orthographic map projection where the perspective points are at infinite distance from the point of tangency of the map surface to the datum surface. The parallel incidence lines are incident from the opposite side of the Earth.

Two well-known non-perspective azimuthal projections are azimuthal equidistant projection and Lambert's equal-area projection.

Table 4.2 lists the applications of important map projections.

4.9 Summary

This chapter on map projection starts with answers to the questions such as 'What is map projection?' and 'Why is it necessary?' It gives an unambiguous mathematical definition of map projection. The process flow of map projection is depicted through a sequence diagram. The three important classes (a) azimuthal projection, (b) cylindrical projection and (c) conical projection are discussed with complete mathematical derivations and diagrams. A generic case of map projection is derived to illustrate the geometrical concepts of map projection. The generic case is further used to derive the special cases of map projections from the perspective of the viewer such as the gnomonic,

Name of the Map Projection	Applications
Cylindrical map projections	Topographical mapping and navigation
Mercator	Display accurate compass bearing for navigation at sea
Transverse Mercator	Preparation of large scale maps of 1:10,000, Pan-Europe mapping at large scale 1:500000 is carried out using transverse Mercator projection
Universal transverse Mercator	Projecting the world in its entirety, for indexing of UTM zones
Cylindrical equidistant	For display of space and aerial images. Google Earth uses the equidistant cylindrical projection (ECP) for displaying its images
Pseudo-cylindrical	For preparation of thematic world mapping, for a more realistic view of the world
Albers equal area conic	For mapping of the US thematic map and for the preparation of world atlases, for management of land resources
Lamberts Conformal Conic (LCC)	For preparation of topographic maps of 1:5,00,0000 scale in the mid-latitude region of the world for planning of land resources
Simple conic	Maps of Russia and Europe are better portrayed using conic projection with two standard parallels
Polyconic	For large-scale mapping of the US and for preparation of coastal charts by US coastal geodetic survey
Gnomonic cylindrical	For flight navigation and preparation of air navigation charts
Azimuthal	
Azimuthal equal area (oblique)	For sea map preparation of the Pacific ocean for hydrocarbon exploration
Azimuthal equal area (polar)	For mapping of the south and north polar regions for polar expeditions
Orthographic (oblique)	For surveillance and recognisance of targets in aerial photo and oblique aerial photography, for creating pictorial view of Earth or its portions
Stereographic (oblique)	For space exploration and for identification of location of space voyage for landing of spacecraft
Stereographic (polar)	For mapping of Arctic and Antarctica regions or polar regions of Earth
Azimuthal equidistant	Large scale mapping of the world
Sinusoidal	For mapping of Moon surface and Mars surface for exploration
Space oblique Mercator	Satellite image mapping

TABLE 4.2

Applications of Map Projections

stereographic and orthographic map projections. Also the inverse map projections are derived for the three major classes of projections. The different types of map projections are classified from the perspective of the plane of projection, cartographic quantities they preserve, the location of the viewer and the way the projection surface touches the datum surface. Finally, to appreciate the use of map projections, a list of applications of map projections is given.

5

Algorithms for Rectification of Geometric Distortions

A large subset of the input domain of GIS constitutes images obtained from different sources and sensors. Images of Earth acquired through satellite or airborne platforms carrying sensors such as RADAR, thermal, infrared (IR), optical, microwave etc. are major input sources of GIS. All these platforms carrying the sensors acquire the image of the Earth's surface corresponding to the IFOV (Instantaneous Field of View) of the sensors. These images have both the signature of the objects and the time stamp of the instance of the image acquired. These images are prone to various errors because of the instability of the platform acquiring the image, imprecise calibration of the sensors or due to radiometric disturbance in the atmosphere in between the sensors and the Earth's surface. Therefore, removal of these errors is important to its further processing by GIS. Also registration of the images precisely with the area of the Earth's surface is an essential prerequisite for effective analysis and measurement in a GIS environment.

In this chapter we give a small survey of research literature on image registration followed by the definition of image registration. Various sources and reasons for these errors are discussed. The steps required for registration of images is discussed. Important algorithms used for registration of satellite images and their analysis are carried out with illustrations. A set of important applications of these registration algorithms is given for a better appreciation of their usage in different domains.

Image registration has been an active area of research for more than three decades. Survey and classification of image registration methods can be found in literature by Zitova et al. [58], Brown [7] and Panigrahi et al.[45]. Separate collections of work covering methods for registration for medical images have been edited by Maintz et al. [34] in a special issue on image registration and vision computing and in a special issue of *IEEE Transactions on Medical Imaging*. A special collection of work covering general methodologies in image registration has been edited by Goshtasby and LeMoigne in a special issue of *Pattern Recognition* [22]. An algorithm for rectification of geometric distortions in satellite images has been illustrated in [45].

5.1   Sources of Geometric Distortion

There are potentially many sources of geometric distortion in remotely sensed images obtained through satellites or airborne platforms. These distortions are more severe and their effects more prominent than radiometric distortions [27]. Listed below are number of factors that may cause geometric distortions in images.

1. The effect of rotation of the Earth during image acquisition.

2. The finite rate of scan of the sensors imaging the Earth's surface.

3. The dimension and geometry of the IFOV of the sensor.

4. The curvature of the Earth at the IFOV.

5. Sensor non-linearity and idealities.

6. Variations in platform altitude and velocity during the image acquisition.

7. Instability of the sensor platform.

8. Panoramic effects related to the imaging geometry.

Zitova and Flusser [58] have investigated a number of algorithms which process and remove geometric distortions in satellite images. Ortho-correction and geometric error correction, are a few important examples where removal of geometric distortions plays a crucial role. These algorithms require supplementary meta-information of the satellite images such as ground control points and correspondence, sensor orientation details, elevation profile of the terrain etc. to establish corresponding transformations. The pre-processing algorithm which removes systematic distortions in the satellite image is discussed. These algorithms which are also known as output-to-input transformations compute the value of the pixels in the input image, for each of the spatial locations in the output image. Efficient methods for implementation of registering image to image were discussed in Pratt [48] and Panigrahi et al. [41]. The transformation computes the coordinate of each output pixel corresponding to the input pixel of an image based on the model established by the Polynomial Affine Transformation (PAT). The transformation is established by the exact amount of scaling, rotation and translation needed for each pixel in the input image so that the distortion induced during the recording stage is corrected.

Following image pre-processing, all images appear as if they were acquired from the same sensor received by remote sensing and represent the geometry and geometric properties on the ground. To detect the change from multi-dated satellite images, geometric error has to be removed first. Geometric error correction becomes especially important when scene to scene comparisons of individual pixels in applications such as change detection are being sought.

In the next section we discuss the prominent registration algorithms used for satellite image registration.

5.1.1 Definition and Terminologies

Image registration is the process of spatially aligning two or more images of the same scene taken from the same or different sensors. This basic capability is needed in various image analysis applications.

Before we discuss different image registration algorithms it is pertinent to define the following terminologies which are commonly used in image registration.

1. **Reference Image:** One of the images in a set of two. This image is kept unchanged and is used as the reference. The reference image is also known as the source image.

2. **Sensed Image:** The second image in a set of two multi-dated images. This image is resampled to register the reference image. The second image is also known as the target image.

3. **Transformation Function:** The function that maps the sensed image to the reference image. Transformation function is determined using the coordinates of the number of correspondence points in the images.

5.1.2 Steps in Image Registration

Given two or more images of the same scene taken at different times or taken by different sensors, the following steps are usually taken to register the images.

Step 1: Image Pre-Processing: This involves preparing the images for feature selection and to establish correspondence between the features using methods such as scale adjustment, noise removal, and segmentation. When pixel sizes in the images to be registered are different but known, one image is resampled to the scale of the other image. This scale adjustment facilitates feature correspondence. If the given images are known to be noisy, they are smoothed to reduce the noise. Image segmentation is the process of partitioning an image into regions so that features can be extracted. The general pre-processing operations used in satellite image registration include image rotation, image smoothing, image sharpening, and image segmentation. There is often a need to pre-process images before registration. Noise reduction and removal of motion blur or haze in images improve feature selection. Image segmentation and edge detection also facilitate feature selection.

Step 2: Feature Selection: Image features are unique image properties that can be used to establish correspondence between two images. The most desired features are pixels or point features, because their coordinates can be directly used to determine the parameters of a transformation function that registers the images. In some images it may not be possible to detect point features; however, lines or regions may be abundant. In such situations points are derived from the lines and regions. For example, the intersections of corresponding line pairs produce corresponding points and centroids of corresponding regions produce corresponding points. To register two images, a number of features are selected from the images and correspondence is established between them. Knowing the correspondences, a transformation function is then found to resample the sensed image to the geometry of the reference image. Generally the features used for image registration are corners, lines, templates, regions, and patches. The signature of objects in a satellite image have structured geometry such as line, square, rectangle, pyramid etc. which are man made objects whereas the natural objects such as mountains, lakes etc. have irregular geometry of contours or irregular polygons. In 3D image, surface patches and regions are often present. Templates are abundant in both 2D and 3D images and are being used as features to register images.

Feature selection is one of the current areas of research in image registration [42]. Many algorithms have been proposed for feature selection. The literature on image registration over the past decade indicates a consistent effort to select features from images which are invariant to sensor geometry and calibration, noise in the atmosphere, intervening lighting condition etc. Hence the effort is to obtain PSRT (Position, Scale, Rotation and Translation) invariant features from the scene. Also, efforts are in place to develop robust and efficient algorithms for feature selection in the scene.

There are many methods for selection of features reported in the literature. Of particular interest are methods involving detection of corners by Harris corner detector [24], line detection by Canny's operator [10], region detection and templates such as Low's SIFT (Scale Invariant Feature Transform) [33], Panigrahi et al. [43].

Step 3: Establishing Feature Correspondence: Having obtained the invariant features in the scene, the next step of image registration is to establish a correspondence between the features of the different scenes. This can be achieved either by selecting features in the reference image and searching for them in the sensed image or by selecting features in both images independently and then determining the correspondence between them. The former method is chosen when the features contain considerable information, such as image regions or templates. The latter method is used when individual features such as points and lines do not contain sufficient information. If the features are not points, it is important that from each pair of corresponding features at least one pair of corresponding points is determined. The coordinates of the corresponding points are used to determine the transformation parameters. For instance, if templates are used, the center of corresponding templates represents corresponding points; if lines are used, the intersection of corresponding line pairs represents corresponding points; and if curves are used, locally maximum curvature points on corresponding curves represent corresponding points. Methods used for establishing correspondence between features are SIFT, PA SIFT, Minimum Mean Square (MMS) difference etc.

Step 4: Determination of Transformation Function: Knowing the coordinates of a set of corresponding points in the images, a transformation function is established to resample the sensed image to the geometry of the reference image. The type of transformation function used should depend on the type of geometric difference between images. If geometric difference between the images is known, in the form of translation, rotation and scaling then the transformation can easily adapt to the geometric difference between the images. Prominent among the transformation functions used to resample sensed satellite image are Log-Polar Transformation and PAT [41].

Step 5: Image Resampling: After establishing the transformation function, the final process of image registration is to apply the transformation function to the sensed image to resample to the geometry of the reference image. This enables fusion of the information in the images or detection of changes in the scenes.

5.2 Algorithms for Satellite Image Registration

Often the images to be registered have scale differences and contain noise, motion blur, haze, and sensor nonlinearities. The size of pixels in terms of ground length or image length are often known and, therefore, either image can be resampled to the scale of the other, or both images can be resampled to a common scale. This resampling facilitates the feature selection and correspondence step of image registration. Depending upon the features to be selected, it may be necessary to segment the images. In this section the prominent satellite image registration methods listed below are discussed.

1. PAT

2. Similarity transformation

3. SIFT

4. Fourier Mellin transform (log-polar transformation)

5.2.1 Polynomial Affine Transformation (PAT)

The equation for a general affine transformation [41] in R^2 is defined by M: $R^2 \to R^2$ and given by a simple equation

$$(k, l) = M(i, j) \tag{5.1}$$

where (i, j) is the coordinate of output image with affine error and (k, l) is the coordinate of input image, obtained at an earlier date say at date (D) in this case. M is the affine transformation, which transforms set of $(i, j) \in R^2$ to set of $(k, l) \in R^2$. In other words, for each pixel (i, j) in the output image, compute its corresponding location (k, l) in input image, obtain the pixel value from input image and put it in output image. Since a reverse computation of pixel location is used, this process is also known as reverse transformation or inverse transformation or output-to-input transformation. The above transformation can be expressed through a pair of polynomials as 5.2 and 5.3

$$k = Q(i, j) = q_0 + q_1 i + q_2 j + q_3 ij \qquad (5.2)$$

and

$$l = R(i, j) = r_0 + r_1 i + r_2 j + r_3 ij \qquad (5.3)$$

These polynomial equations can be represented in matrix form. Since the affine transformation is represented through a set of polynomials, it is called PAT. The unknown coefficients q_i and r_i are obtained after solving the following system matrix representing polynomial.

$$K = MQ \qquad (5.4)$$

and

$$Q = M^{-1}K \qquad (5.5)$$

The dimension and condition of the system matrix M depends upon the number of GCP-CP pairs selected and their spatial distribution in the image. At least 3 pairs of non-collinear GCP need to be selected to establish an affine frame in R^2. If 3 pairs of GCP are selected then we get a 3x3 square matrix representing the transformation, which can solve the translation, rotation and scaling distortions in the input image. If the GCPs are collinear and densely populated then the matrix is ill-conditioned and sometimes it leads to inconsistency and rank deficiency. Hence choice of more numbers of GCP-CP pairs leads to removal of highly irregular geometric distortions and to greater accuracy. Least Squares Method (LSM) is used to avoid inconsistency and solve the above matrix equation for robust results. By definition, LSM is the one that minimizes

$||K - MQ||^2$, which when solved leads to 5.6

$$Q_{LSM} = [M^T M]^{-1} M^T K \qquad (5.6)$$

Similarly matrix equation for second polynomial can be derived and solved for R resulting in 5.7

$$R_{LSM} = [M^T M]^{-1} M^T L \qquad (5.7)$$

5.2.2   Similarity Transformation

The similarity transformation or the transformation of the Cartesian coordinate system represents global translation, rotation, and scaling difference between two images. This is defined by the pair of equations 5.8a and 5.8b

$$X = S[x \cos \theta + y \sin \theta] + h \tag{5.8a}$$

and

$$Y = S[-x \sin \theta + y \cos \theta] + k \tag{5.8b}$$

where S, θ and (h, k) are scaling, rotational, and translational differences between the images respectively. These four parameters can be determined if the coordinates of two corresponding points in the images are known. The rotational difference is computed from the angle between the lines connecting the two points in the image. The scaling difference between the images is determined from the the of the distances between the images. Knowing the scaling and rotation the translation parameters (h, k) are determined by substituting the coordinates of midpoints of the lines connecting the points into the above equations and solving for h and k. Therefore similarity transformation can be applied to image to image registration where, the exact amount or scaling, rotation and translation of the image to be registered is known.

5.3 Scale Invariant Feature Transform (SIFT)

Automatic extraction of key point, dominant point [42], Ground Correlation Point (GCP) [55] or Control Points (CP) from images in general and satellite images in particular is an active area of research. Extraction of key point set from satellite images is a precursor to establishing correspondence between images to be registered. Once correspondence is established, the transformation can be established for registration. A good key point in an image is one which is invariant to radiometric and geometric disturbance in the image. Considerable research in computer vision has been carried out to extract key points from an image. SIFT proposed by Lowe [33] extracts fairly stable key features in the image which are invariant under geometric, radiometric and illumination variations. The features are in variant to image scale and rotation, and are shown to provide robust matching across a substantial range of affine distortion, change in 3D viewpoint, addition of noise, and change in illumination. The features are highly distinctive, in the sense that a single feature can be correctly matched with high probability against a large database of features from many images.

The following are the major computation steps used to generate the set of image features which are scale invariant:

Detection of Scale-Space Extrema: The first stage of computation searches overall scales and image locations. It is implemented efficiently by using a difference-of-Gaussian function to identify potential interest points that are invariant to scale and orientation.

Localization of Key Points: At each candidate location, a detailed model is fit to determine location and scale. Key points are selected based on measures of their stability.

Orientation Assignment: One or more orientations are assigned to each keypoint location based on local image gradient directions. All future operations are performed on image data that has been transformed relative to the assigned orientation, scale, and location for each feature, thereby providing invariance to these transformations.

Key Point Descriptor: The local image gradients are measured at the selected scale in the region around each key point. These are transformed into a representation that allows for significant levels of local shape distortion and change in illumination.

5.3.1   Detection of Scale-Space Extrema

Key points are detected using a cascade filtering approach. The first stage of key point detection is to identify locations and scales that can be repeatably assigned under differing views of the same object. Detecting locations that are invariant to scale change of the image is accomplished by searching for stable features across all possible scales, using a continuous function of scale known as scale space as depicted in Figure 5.2. It has been shown by Koenderink and Lindeberg [28], [29] that under a variety of reasonable assumptions the only possible scale-space kernel is the Gaussian function. Therefore, the scale space of an image is defined as a function, $L(x, y, \sigma)$, that is produced from the convolution of a variable-scale Gaussian, $G(x, y, \sigma)$, with an input image, $I(x, y)$:

$$L(x, y, \sigma) = G(x, y, \sigma) * I(x, y) \tag{5.9}$$

where '*' is the convolution operation in 'x' and 'y' and

$$G(x, y, \sigma) = \frac{1}{2\pi\sigma^2} \exp^{-\frac{x^2+y^2}{2\sigma^2}} \tag{5.10}$$

To efficiently detect stable key point locations in scale space, Lowe [33] proposed searching for key points using scale-space extrema. Where the scale space is simulated using the Difference-of-Gaussian (DoG) function convolved with the image, $D(x, y, \sigma)$. DoG is computed from the difference of two images from nearby scales separated by a constant multiplicative factor k:

$$D(x, y, \sigma) = (G(x, y, k\sigma) - G(x, y, (k-1)\sigma)) * I(x, y) \tag{5.11}$$

5.3.2   Local Extrema Detection

In order to detect the local maxima and minima of $D(x, y, \sigma)$, the intensity value of each sample point is compared to its eight neighbours in the current image and nine neighbours in the scale above and below. It is selected only if its intensity is larger than all of these neighbours or smaller than all of them. The process of computing SIFT features from a satellite image is depicted in Figure 5.1

5.3.3 Accurate Key Point Localization

Once a key point candidate has been found by comparing a pixel to its neighbours, the next step is to perform a detailed fit to the nearby data for location, scale, and ratio of principal curvatures. This information allows points to be rejected that have low contrast (and are therefore sensitive to noise) or are poorly localized along an edge. The initial implementation of this approach by Lowe [33] simply located key points in images at various scales.

Figure 5.2 depicts the scale-space pyramid representation as framework for multi-scale signal representation. Scale-space theory is a framework for multi-scale signal representation developed by the computer vision, image processing and signal processing communities with complementary motivations from physics and biological vision. It is a formal theory for handling image structure points, at different scales, by representing an image as a one-parameter family of smoothed images, the scale-space representation, parameterized by the size of the smoothing kernel used for suppressing fine-scale structures. The parameter 'σ' in this family is referred to as the scale parameter, with the interpretation that image structures of spatial size smaller than about have largely been smoothed away in the scale-space level at scale 'σ'. The main type of scale space is the linear (Gaussian) scale space, which has wide applicability as well as the attractive property of being possible to derive from a small set of scale-space axioms.

However, recently Lowe [32] has developed a method for fitting a 3D quadratic function to the local sample points to determine the interpolated location of the maximum, and his experiments showed that this provides a substantial improvement to matching and stability. This approach uses the Taylor expansion up to the quadratic terms) of the scale-space function, $D(x, y, \sigma)$, shifted so that the origin is at the sample point.

$$D(x) = D + \frac{\partial D^T}{\partial x} x + \frac{1}{2} x^T \frac{\partial^2 D}{\partial x^2} x \qquad (5.12)$$

where 'D' and its derivatives are evaluated at the sample point and $x = (x, y, \sigma)^T$ is the offset from this point. The location of the extremum, 'x', is determined by taking the derivative of this function with respect to x and setting it to zero, giving

$$X' = -\frac{\partial^2 D^{-1}}{\partial x^2} \frac{\partial D}{\partial x} \qquad (5.13)$$

In Figure 5.3, the image is convolved with Gaussian function. Due to the convolution with Gaussian function the high frequency component is removed.

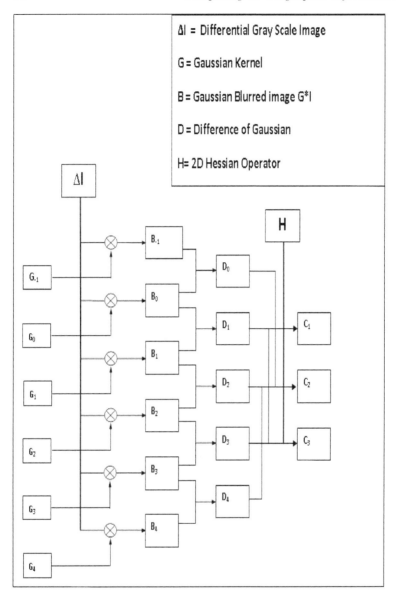

FIGURE 5.1

Steps of computing key points from satellite image using SIFT, detection of key points form image using DOG and maximization rule. The image is convolved with Gaussian function. Due to this the high frequency noise is removed from the image and the image become blurred. The high frequency corresponds to minute image details that are never common between two different images so these are removed. The blurred images are subsequently subtracted from each other to again remove details in two subsequent blurs. Now using maximization rule a pixel which has the highest magnitude of all its neighbours is selected as key point.

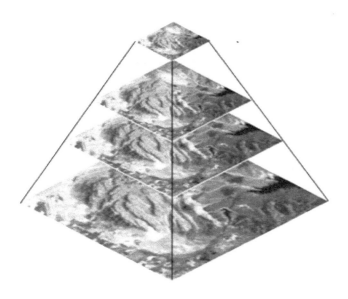

FIGURE 5.2
Gaussian blurred image pyramid, depicting the scale space of an image

The Gaussian convolved images appear blurred in comparison to the original image simulating the scale-space visualization of the image. The high frequency corresponds to minute image details that are never common between two different images so these are removed. The blurred images are subsequently subtracted from each other to again remove details in two subsequent blurs. These successive blurred images in each stage of convolution are designated in Figure 5.3 as B_i and D_i. Using maximization rule [33] a pixel which has the highest magnitude of all its neighbours in the present scale, preceeding scale and succeeding scale is selected as the key point.

As suggested by Lowe [33], the hessian and derivative of $D(x)$ which is the difference of Gaussian at x, are approximated by using differences of intensity values of the neighbouring sample points. The resulting 3x3 linear system can be solved with minimal cost. If the offset 'x' is larger than 0.5 in any dimension, then it means that the extremum lies closer to a different sample point. In this case, the sample point is changed and the interpolation performed instead about that point. The final offset 'x' is added to the location of its sample point to get the interpolated estimate for the location of the extremum. The function value at the extremum, D(x'), is useful for rejecting unstable extrema with low contrast. This can be obtained by substituting equation 5.12 into 5.11, giving

$$Dx^{'} = D(x') + 0.5\frac{\partial D^T}{\partial x}x^{'} \qquad (5.14)$$

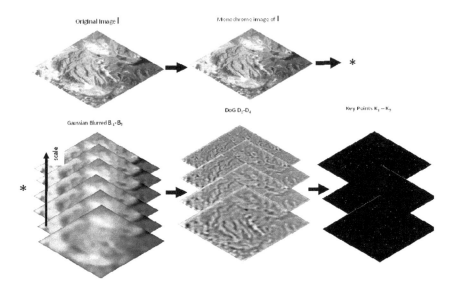

FIGURE 5.3
Detection of keypoint from image using DoG and maximization rule.
The D_0 to D_4 are the Difference of Gaussian (DoG) images obtained after computing the difference of successive Gaussian blurred images. K_1 to K_3 are the set of key features of the image computed using SIFT.

From the experiments that have been carried out, all extrema with a value of $|D(x')|$ less than 0.03 were discarded.

5.3.4   Eliminating Edge Responses

For stability, it is not sufficient to reject key points with low contrast. The DoG function will have a strong response along edges, even if the location along the edge is poorly determined and therefore unstable to small amounts of noise. A poorly defined peak in the DoG function will have a large principal curvature across the edge but a small one in the perpendicular direction. The principal curvatures can be computed from a 2x2 Hessian matrix, H, computed at the location and scale of the key point

$$H(x,y) = \begin{pmatrix} \frac{\partial^2 I}{\partial x^2} & \frac{\partial^2 I}{\partial x \partial y} \\ \frac{\partial^2 I}{\partial y \partial x} & \frac{\partial^2 I}{\partial y^2} \end{pmatrix} \tag{5.15}$$

The derivatives are estimated by taking differences of neighbouring sample points. The eigenvalues of H are proportional to the principal curvatures of D. Borrowing from the approach used by Harris and Stephens [24], we can

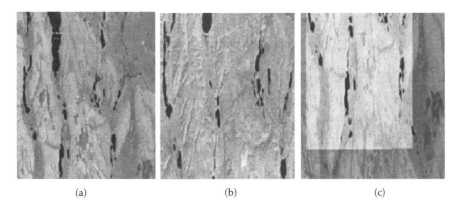

(a)   (b)   (c)

FIGURE 5.4
Example of registration of satellite image pair using Log-Polar transformation:
(a) base image, (b) image with geometric error, (c) image (b) registered and
resampled with respect to image (a)

avoid explicit computing of the eigenvalues, as of H and their product from
the determinant:

$$Tr(H) = D_{xx} + D_{yy} \tag{5.16}$$

$$Det(H) = D_{xx}D_{yy} - D_{xy}^2 \tag{5.17}$$

In the unlikely event that the determinant is negative, the curvatures have
different signs so the point is discarded as not being an extremum. Let 'r' be
the ratio between the largest magnitude eigenvalue and the smaller one, so that
$\lambda_{max} = r\lambda_{min}$. Then 'r' depends only on the ratio of the eigenvalues rather
than their individual values. The quantity $\frac{2(r+1)}{r}$ is at a minimum when the
two eigenvalues are equal and it increases with 'r'. To eliminate edge response
we need to check the key points which meet the criteria.

$$\frac{Tr(H)^2}{Det(H)} \leq \frac{(r+1)^2}{r^2} \tag{5.18}$$

If we set $r = 10$ then the key points that have ratio of principal curvature
greater than 10 are eliminated from the list of key points thus eliminating the
key points which are in the edge of an image and are considered as highly
unstable.

Finally the SIFT features computed using this method is used for establishing correspondence between multi-dated satellite image for registration as
depicted in Figure 5.5.

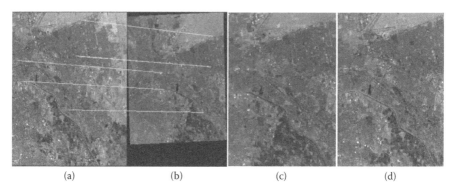

(a)   (b)   (c)   (d)

FIGURE 5.5
Satellite images: (a) base image, (b) image with geometric distortion, (c) image, (b) registered with respect to image (a), (d) final registered image (b)

5.4 Fourier Mellin Transform

The Fourier Transform (FT) is an operation that transforms one complex-valued function of a real variable into another. In applications such as signal processing, the signal is a function of time domain. The transformed function is of frequency domain, therefore called the frequency domain representation of the time domain function. The FT and the inverse transformation are given by following equations respectively

$$F(\hat{\xi}) = \int f(x)e^{-2\pi I x\xi}\, dx \qquad (5.19)$$

and the inverse FT

$$F(x) = \int f(\hat{x})e^{2\pi I x\xi}\, d\xi \qquad (5.20)$$

For automatic registration of spatial images, the FT-based method makes use of the Fourier shift theorem, which guarantees that the phase of a specially defined 'ratio' is equal to the phase difference between the images. It is known that if two images I_1 and I_2 differ only by a shift, (x_0, y_0), i.e., $I_2(x, y) = I_1(x - x_0, y - y_0)$, then their FTs are related by the formula

$$F_2(\xi, \eta) = e^{-j2*\pi*(\xi.x_0 + \eta*y_0)} * F_1(\xi, \eta) \qquad (5.21)$$

The ratio of two images I_1 and I_2 is defined as:

$$R = \frac{F_1(\xi, \eta) * conj(F_2(\xi, \eta))}{|(F_1(\xi, \eta))| * |(F_2(\xi, \eta))|} \qquad (5.22)$$

where conj is the complex conjugate, and abs is absolute value. By taking the inverse FT of R, we see that the resulting function is approximately zero everywhere except for a small neighbourhood around a single point. This single point is where the absolute value of the inverse Fourier transfer of R attains its maximum value. It can be shown that the location of this point is exactly the displacement (x_0, y_0) needed to optimally register the images. If the two images differ by shift, rotation and scaling, then converting abs (F) from rectangular coordinates (x, y) to log-polar coordinates makes it possible to represent both rotation and scaling as shifts. The Fourier Mellin transform is a useful mathematical tool for image recognition because its resulting spectrum is invariant in rotation, translation and scaling. The Fourier transform itself is translation invariant and its conversion to log-polar coordinates converts the scale and rotation differences to vertical and horizontal offsets that can be measured. A second FT, called the Mellin transform (MT) gives a transform-space image that is invariant to translation, rotation and scale.

$$M(u, v) = \int \int f(x, y) x^{-ju-1} y^{-jv-1} dx dy \tag{5.23}$$

FT-based automatic registration relies on the Fourier shift theorem, which guarantees that the phase of a specifically defined 'ratio' is equal to the phase difference between the images. To find displacement between two images, compute the ratio $R = F_1 conj(F_2)/|F_1 F_2|)$ and apply inverse Fourier transform. By applying the inverse Fourier transform to the ratio, we get an array of numbers returned that is zero everywhere except for a small area around a single point. By using the Max function one can find the maximum value. The location of the Max value is exactly the displacement (x_0, y_0) that is needed to optimally register the images.

5.4.1 The Log-Polar Transformation Algorithm

The following computing steps describe the log-polar transformation which when applied to a pair of unregistered satellite images, register the second image with geometric distortions to the first image. In this the first image is treated as the base image and the second image is considered as the currently obtained image to be registered with respect to the base image.

1. Read in I_1 - the base image to register against
2. Read in I_2 - the image to be registered
3. Take the FFT of I_1, shifting it to center on zero frequency
4. Take the FFT of I_2, shifting it to center on zero frequency
5. Convolve the magnitude of (3) with a high pass filter
6. Convolve the magnitude of (4) with a high pass filter

7. Transform (5) into log polar space

8. Transform (6) into log polar space

9. Take the FFT of (7)

10. Take the FFT of (8)

11. Compute phase correlation of (9) and (10)

12. Find the location (x, y) in (11) of the peak of the phase correlation

13. Compute angle (360 / Image Y Size) * y from (12)

14. Rotate the image from (2) by - angle from (13)

15. Rotate the image from (2) by - angle + 180 from (13)

16. Take the FFT of (14)

17. Take the FFT of (15)

18. Compute phase correlation of (3) and (16)

19. Compute phase correlation of (3) and (17)

20. Find the location (x,y) in (18) of the peak of the phase correlation

21. Find the location (x,y) in (19) of the peak of the phase correlation

22. If phase peak in (20) $\geq$ phase peak in (21), (y,x) from (20) is the translation

23. (a) Else (y,x) from (21) is the translation and also:

24. (b) If the angle from (13) $\leq$ 180, add 180 to it, else subtract 180 from it.

The above method is applied to a pair of satellite image and the results are displayed in the Figure 5.5.

5.5   Multiresolution Image Analysis

Multiresolution image analysis through multi-scale approximation of image is an important technique for analyzing an image in different scale. The contents of an image such as pixels, edges and objects can be analyzed in different scale and summed up to study the overall analysis of the image contents. This image processing method makes use of different techniques such as Discrete Wavelet Transforms (DWT), Laplacian of Gaussian (LoG), Difference of Gaussian (DoG) etc. to simulate the scale-space representation of an image. The multi-resolution technique has its utilization in many signal processing applications in general and image processing applications in particular. The main idea of image pyramid or 'pyramid representation' of an image is a type of multi-scale signal representation developed by the computer vision, image processing and signal processing communities, in which a signal or an image is subject to repeated smoothing and subsampling. Historically, pyramid representation is a predecessor to scale-space representation and multiresolution analysis.

The first multi-resolution technique was introduced by Stephane Mallat and Yves Meyer in 1988-89 [36]. Witkin made use of the theory for scale-space filtering of signals [56]. The microlocal analysis of image structures using image pyramid and difference equations was introduced by Peter J. Burt et. al., in 1981-83 [9].

There are two main types of image pyramid: lowpass pyramids and bandpass pyramids. A lowpass pyramid is generated by first smoothing the image with an appropriate smoothing filter and then subsampling the smoothed image, usually by a factor of two along each coordinate direction [1]. This smoothed image is then subjected to the same processing, resulting in a yet smaller image. As this process proceeds, the result will be a set of gradually more smoothed images, where the spatial sampling density decreases level by level. If illustrated graphically, this multiscale representation will look like a pyramid, from which the name has been obtained. A bandpass pyramid is obtained by forming the difference between adjacent levels in a pyramid, where in addition to sampling some kind of interpolation is performed between representations at adjacent levels of resolution, to enable the computation of pixel-wise differences.

A variety of different smoothing kernels have been proposed for generating pyramids. Gaussian, Laplacian of Gaussian and Wavelet are some of the highly used smoothing kernel in an image processing community [1], [50], [18]. In the early days of computer vision, pyramids were used as the main type of multi-scale representation for computing multi-scale image features from real-world image data [37]. More recent techniques include scale-space representation, which has been popular among some researchers due to its theoretical foundation, the ability to decouple the subsampling stage from the multi-scale representation, the more powerful tools for theoretical analysis as well as the ability to compute a representation at any desired scale, thus avoiding the algorithmic problems of relating image representations at different resolution. Nevertheless, pyramids are still frequently used for expressing computationally efficient approximations to scale-space representation.

5.6   Applications of Image Registration

Image registration in general and satellite image registration in particular finds numerous applications in the domain of Digital Image Processing (DIP), Medical Image Processing (MIP), Spatial Data Mining and Computer Vision

Name of the Registration Algorithm	Areas of Applications
Polynomial Affine Transformation (PAT)	Satellite image registration with topographic vector map. Registration of multi-dated satellite image for image change detection and Land use land cover change detection.
Similarity Transformation (ST)	Registration of medical image for change detection analysis of ocular deformation and malign tumor detection.
Scale Invariant Feature Detection (SIFT)	Registration of images with varying scale and lighting conditions. For multi-sensor data fusion and computer vision applications.
Log-Polar Transformation	For registration of medical image in data mining applications such as detection of cancerous cells in the MRI (Magnetic Resonance Imaging)

TABLE 5.1

Applications of Image Registration Algorithms

(CV), GIS. Image registration is an essential prerequisite for image change detection which has a vast array of applications in diverse domains. The table below gives a candidate application of image registration algorithms.

Applications of these image registration techniques in different domains are given for a better appreciation their usage 5.6. Image registration is very important for fusion of images from same sensor or from different sensors so that the pixel level correlation of the images is obtained. Image fusion is used for sharpening of defocused images or restoration of noisy images. Also to enhance images radiometric and spatial resolution image fusion is used which in turn depends on highly accurate image registration techniques. Registration of medical images obtained using PET or MRI of the similar area of the body in a controlled environment, Similarity transformation is used for registration. Medical image registration is a prerequisite for detection of change. Generally the changes are detected in retinal cell of the eye ball or detection of tumor or cancerous cells from medical images. Change detection using multi-dated satellite images finds application in many domains. One such application is for land use and land cover analysis using remotely sensed images. Because the images used are obtained using same satellite, PAT is the best used for registration of these multi-dated satellite images. Detection of objects of operational interest from satellite images using different sensors or images taken with different radiometric and illumination conditions are useful for extraction of terrain intelligence. Since the images to be registered or compared are of different scale and illumination condition, SIFT is the best prescribed method for feature extraction and registration.

5.7 Summary

This chapter presents a brief review of research literatures reporting the image registration techniques. The various sources of errors in satellite image are discussed. The definition of image registration followed with the basic concepts and taxonomy of image registration techniques are given for better appreciation of the computing steps involved. The generic steps involved in registration of images are discussed. Important image registration techniques such as Polynomial Affine Transformation (PAT), Similarity Transform, Scale Invariant Feature Transform (SIFT), Fourier Mellin Transformation and Log-Polar Transformation were discussed in detail. From the analysis of these registration algorithms it can be easily concluded that if the translation, rotation and scaling parameters of the image with respect to the image to be registered are known then similarity transformation is to be used to achieve the registration. This registration process is the easiest and highly accurate. If none of the parameters of the images to be registered are known and few distinguished features are common between the images then PAT is the best method for registration. PAT achieves relatively low accuracy registration in comparison to similarity transformation. If the images to be registered are obtained in different lighting and radiometric conditions SIFT is the best to select the similar features and establishing correspondence between them before registration. This is a highly robust and efficient technique which used scale space properties of the image for selection of key feature in the image for correspondence. If the images are taken in a controlled environment such as medical images then Fourier Mellin Transformation and Log-Polar Transform are the best techniques for registration.

6

Differential Geometric Principles and Operators

Differential geometry constitutes of a set of mathematical methods and operators which are useful in computing geometric quantities of discrete continuous structures using differential calculus as discussed by Koenderink [28],[29]. It can be described as a mathematical tool box for describing shape through derivatives. The assumption made in differential geometry is that the geometrical structures such as curves, surfaces, lattices etc. are everywhere differentiable and there are no sharp discontinuities such as corners or cuts. In differential geometry of shape measures, there is no global coordinate system. All measurements are made relative to the local tangent plane or normal. Although image is spatial data, there is no coordinate system associated with an image. In case of a geo-coded, geo-referenced satellite image, a global coordinate system such as UTM (Universal Transverse Mercator) is implicitly associated. Also for the sake of referencing, the positional value of each pixel in an image it is often associated with a coordinate system that defines the position of the origin. The intensity of the images is treated as perpendicular to the image plane or the 'Z' coordinate of the image making it a 3D surface. Further the application of pure geometry in GIS is discussed by Brannan et al. [6].

The following geometric quantities can be computed from the intensity profile corresponding to an image and analyzed using differential geometry.

1. Gradient
2. Curvature
3. Hessian
4. Principal curvature, Gaussian curvature and mean curvature
5. Laplacian

6.1 Gradient (First Derivative)

Gradient is the primary first-order differential quantity of a surface. For the intensity profile of an image, gradient at pixel location can be computed. The gradient of a function $f(x, y)$ is defined by Equation 6.1.

$$\bigtriangledown f(x, y) = \begin{pmatrix} \frac{df(x,y)}{dx} \\ \frac{df(x,y)}{dy} \end{pmatrix} \tag{6.1}$$

Gradient $\bigtriangledown f$ is a 2D vector quantity. It has both direction and magnitude which vary at every point. Following are the properties of gradient or inferences that can be concluded from the gradient:

1. The gradient direction at a point is the direction of the steepest ascent/descent that point.

2. The gradient magnitude is the steepness of that ascent/descent

3. The gradient direction is the normal to the level curve at that point.

4. The gradient defines the tangent plane at that point.

Hence the gradient can be made use of to compute universal first order information about the change of gradient which is akin to spectral undulation in the terrain surface at the point of an image. One of the fundamental concepts of differential geometry is that, we can describe local surface properties with respect to the coordinate system dictated by the local surface which in our case is an image surface. It is not computed with respect to a global frame of reference or coordinate system. This concept is known as 'gauge coordinate', which is a coordinate system that the surface carries along with itself wherever it goes. This is apt for image surfaces which are being analyzed independent of the global coordinate system in a display device, hard copy or projected screen. Gradient direction is one of the intrinsic properties of the image, independent of the choice of spatial coordinate axis. The gradient direction and its perpendicular constitute the first order gauge coordinates, which are best understood in terms of the images level. An iso-photo is a curve of constant intensity. The normal to the iso-photo curve is the gradient direction of the image.

6.2   Concept of Curvature

In mathematics, curvature refers to a number of loosely related concepts in different areas of geometry. Intuitively, curvature is the amount by which a geometric object deviates from being flat, or straight in the case of a line. Curvature is defined in different ways depending on the context. There is a key distinction between extrinsic curvature and intrinsic curvature. Extrinsic curvature is defined for objects embedded in Euclidean space in a way that relates to the radius of curvature of circles that touch the object. Intrinsic curvature of an object is defined at each point in a Riemannian manifold. The canonical example of extrinsic curvature is that of a circle, which everywhere has curvature equal to the reciprocal of its radius. Smaller circles bend more sharply, and hence have higher curvature. The curvature of a smooth curve is defined as the curvature of its osculating circle at each point.

In a plane, this is a scalar quantity, but in three or more dimensions it is described by a curvature vector that takes into account the direction of the bend as well as its sharpness. The curvature of more complex objects (such as surfaces or even curved n-dimensional spaces) is described by more complex objects from linear algebra, such as the general Riemannian curvature tensor.

From differential geometric point of view curvature is the second order derived geometric quantity of a curve or surface. Curvature is the simplest form of expressing the magnitude or rate of change of gradient at a point in the surface. The curvature which usually is used in calculus is the extrinsic curvature.

In 2D plane let the optical profile be defined by the parametric equation $x = x(t)$ and $y = y(t)$. Then the curvature 'K' is defined by equation 6.2

$$\frac{d\Phi}{dt} = \frac{\frac{d\Phi}{ds}}{\frac{ds}{dt}} = \frac{\frac{d\Phi}{ds}}{\sqrt{(\frac{dx}{dt})^2 + (\frac{dy}{dt})^2}} \tag{6.2}$$

where Φ is the tangent angle to the surface and 't' is the arc length of the surface. Curvature has the unit of inverse distance. The derivative of the numerator of the above equation can be derived using the identity

$$tan\Phi = \frac{dy}{dx} = \frac{\frac{dy}{dt}}{\frac{dx}{dt}} = \frac{y'}{x'} \tag{6.3}$$

Therefore,

$$\frac{d(tan\Phi)}{dt} = Sec^2\Phi \frac{d\Phi}{dt} = \frac{x' y'' - y' x''}{x'^2} \tag{6.4}$$

$$\frac{d\Phi}{dt} = \frac{1}{Sec^2\Phi} \frac{d(tan\Phi)}{dt} = \frac{1}{1 + tan^2\Phi} \frac{x' y'' - y' x''}{x'^2} = \frac{x' y'' - y' x''}{x'^2 + y'^2} \tag{6.5}$$

Combining equations 6.4 and 6.5 the curvature can be computed using the differentiale equation 6.6

$$k = \frac{x' y'' - y' x''}{\sqrt[3]{x'^2 + y'^2}} \tag{6.6}$$

The curvature derived in the above equation 6.6 is for a parametric surface. The curvature in this is known as principal curvature and is usually computed where the surface is continuous and differential and double differential of the surface function is easily computable. For a one-dimensional curve function given by $y = f(x)$ the above formula reduces to

$$k = \frac{\frac{d^2y}{dx^2}}{\sqrt[3]{(1 + (\frac{dy}{dx})^2)}} \tag{6.7}$$

For normal terrain surface where the change in the elevation is not rapid with respect to the displacement in the plane the slope is varying gradually. Hence one can assume the $dy/dx < 1$ and hence the square of the gradient is far less than one making the denominator of the above equation approximately equals to unity. This assumption may not hold true for highly undulated terrain surface.

$$k = \frac{d^2 y}{dx^2} \tag{6.8}$$

In the case of a satellite image which represents the optical profile of the terrain surface, the terrain surface is a grid of pixel values. This is a discrete representation of the continuous surface in the form of a matrix. The pixel values are discrete surrounded by eight pixels in all cardinal directions, except for the pixels in the boundary of the image. To compute the curvature of such surfaces at any grid point, the above formula given in equation 6.8 is used in a modified manner known as Hessian.

An approximate computation of curvature of a continuous surface can be obtained from the second order differential of the surface. This concept is extended to a discrete surface represented in the form of a matrix of values and is implemented through Hessian. Hence curvature of the optical profile of an image given in the form of a matrix is computed using the Hessian matrix which is discussed in the next section.

6.3   Hessian: The Second Order Derivative

The second-order derivative of a surface gives the rate of change of the gradient in the surface which is often connoted as the curvature. This is computed using the matrix of the second order derivatives which is known as the Hessian. An image can be mathematically modeled to be a surface in 2D given by $I = f(x, y)$, where I is the value of intensity which is a function of the spatial location (x, y) in the image plane. The Hessian of such a surface is given by equation 6.9:

$$H(x, y) = \begin{pmatrix} \frac{\partial^2 I}{\partial x^2} & \frac{\partial^2 I}{\partial x \partial y} \\ \frac{\partial^2 I}{\partial y \partial x} & \frac{\partial^2 I}{\partial y^2} \end{pmatrix} \tag{6.9}$$

As the image is not associated with any coordinate system, the ordering of the pixel index can be considered from any arbitrary origin of the image. Therefore

$$\frac{\partial^2 I}{\partial x \partial y} = \frac{\partial^2 I}{\partial y \partial x} \tag{6.10}$$

Also, the value of intensity in an image is always positive and definite. There is no negative intensity value in the image profile. These two make the

Hessian matrix a real, symmetric matrix. One can use Hessian to calculate the second order derivatives in any direction because Hessian is a real and symmetric matrix having the following mathematical properties.

- Its determinant is equal to the product of its eigenvalues and is invariant to the selection of the spatial coordinate (x, y).
- The trace of Hessian matrix Tr(H) (i.e. the sum of the diagonal elements) is also invariant to selection of x and y.

The eigenvalues and eigenvectors of Hessian matrix have great significance which is exploited in the next section to study the geometrical topology of the surface. Eigenvalues play a crucial role to classify the changed pixels into 2D (planar) change or 3D (curved) change category. The physical significances of the eigenvalues are:

1. The first eigenvector (corresponding to the higher eigenvalue) is the direction of the surface curvature with maximum magnitude

2. The second eigenvector (corresponding to the smaller eigenvalue) has the smallest magnitude) is the direction of least curvature in the surface.

3. The magnitude of curvature in the surface is proportional to the magnitude of the eigenvalues.

The eigenvalues of 'H' are called the principal direction of pure curvature and they are always orthogonal. The eigenvalues of Hessian are also called the principal curvature and are invariant under rotation. The principal curvatures are denoted as λ_1 and λ_2 and are always real valued. Principal curvature of a surface is the intrinsic property of the surface. This means the direction and magnitude of principal curvature are independent of the embedding of the surface in any frame of reference or coordinate system.

6.4   Gaussian Curvature

Gaussian curvature is the determinant of Hessian matrix 'H' which is equal to the product of principal curvatures λ_1 and λ_2. Gaussian curvature is denoted as 'K', and is computed by equation 6.11.

$$K = Det(H) = (\lambda_1 * \lambda_2) \tag{6.11}$$

The physical significance of Gaussian curvature can be interpreted as the undulated amount of the surface of the terrain which is in excess to its perimeter. In other words, if one has to make a surface flat, the excess surface with respect to its planimetric area held by the perimeter of the surface has to be removed. This excess surface area is equivalent to the Gaussian curvature.

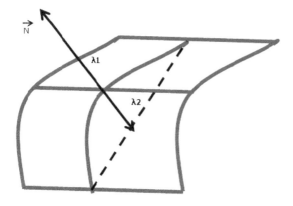

FIGURE 6.1
Edge surface with Gaussian curvature $K = 0, \lambda_1 = 0$ and $\lambda_2 < 0$. The principal eigenvalues are directed in orthogonal directions.

The classification of the image surface can be done using the mean gaussain curvature. For elliptic surface patches the curvature is positive in any directions i.e. $K \geq 0$.

In other words if $H \geq 0$ then the surface is convex.

If $H \leq 0$ then the surface is concave and curvature in any direction is negative.

For hyperbolic patches: $K \leq 0$, the curvature is positive in some direction and negative in some other direction.

For $K = 0$, i.e. one or both of the principal curvature is zero, the surfaces are known as parabolic curved surfaces and they lie in the boundary of elliptic and hyperbolic regions.

Surfaces where $\lambda_1 = \lambda_2$ have principal curvature is same in all directions then the surface can be categorized as planar surface or smooth surface. Often such smooth surfaces are known as 'umbilics'.

Surfaces where $\lambda_1 = -\lambda_2$ points having principal curvature same in magnitude but opposite-sign are known as minimal points. The different types of surfaces according to the relative value of Gaussian curvature and surface normal are explained pictorially in Figures 6.1 to 6.3.

6.5   Mean Curvature

Mean curvature is the average of the principal curvatures λ_1 and λ_2 . It is equivalent to half of the trace of Hessian matrix. The mean curvature is in-

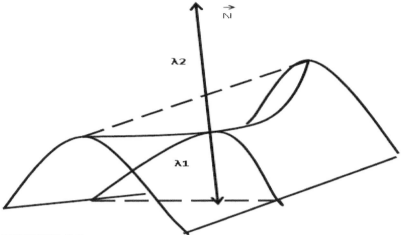

FIGURE 6.2
Saddle surface with Gaussian curvature $K < 0, \lambda_1 < 0$ and $\lambda_2 > 0$ The, principal eigenvalues directed in orthogonal directions of the dominant curvatures

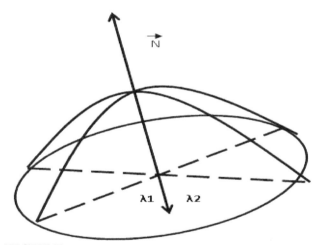

FIGURE 6.3
Blob-like surface with Gaussian curvature $K > 0, \lambda_1 < 0$ and $\lambda_2 < 0$, a convex surface

variant to the selection of x and y making it an intrinsic property of the surface. The mean curvature is given by equation 6.12.

$$H = (\lambda_1 + \lambda_2)/2 \tag{6.12}$$

6.6 The Laplacian

The Laplacian is simply twice the mean curvature and is equivalent to the trace of H. It is also invariant to the rotation of the plane. The Laplacian is given by equation 6.13.

$$Trace(H) = (\lambda_1 + \lambda_2) \tag{6.13}$$

The measure of undulation of a terrain surface which can be proportional to its mean curvature can be computed using equation 6.14. To understand the concept of the deviation from flatness of a terrain surface the amount of the undulated terrain in the surface can be computed using the eigenvalues as given in equation 6.14.

$$Undulation = \lambda_1^2 + \lambda_2^2 \tag{6.14}$$

Curvatures play an important role in study of shape of the terrain surface [28],[29] for 'shape classification' in the coordinate system spanned by the principal eigenvectors $\lambda_1 X \lambda_2$. However the shape descriptors are better explained through its polar coordinate systems as given in equations 6.15 and 6.16 as shape angle and degree of curvature respectively.

$$S = tan^{-1} \frac{\lambda_1}{\lambda_2} \tag{6.15}$$

$$C = \sqrt{(\lambda_1^2 + \lambda_2^2)} \tag{6.16}$$

Degree of curvature is the square root of the deviation from flatness. Hence image points with same S and having different C values can be thought of as being the same shape with different stretch or scale.

6.7 Properties of Gaussian, Hessian and Difference of Gaussian

Properties of Gaussian and Hessian functions poses important mathematical properties to characterize profiles of geometric objects in general and topology of the intensity profiles of the image in particular. Hence we discuss the characteristics and mathematical properties of these functions manifested as a window operator while processing images. What they yield when applied to spatial data in the form of two dimensional matrix is quite interesting.

6.7.1 Gaussian Function

The 2D Gaussian function is given in equation 6.17

$$G(x,y) = \frac{1}{2\pi\sigma^2} * \exp^{\frac{x^2+y^2}{2\sigma^2}} \tag{6.17}$$

where x is the distance from the origin in the horizontal axis, y is the distance from the origin in the vertical axis, and σ is the standard deviation of the Gaussian distribution. When applied in two dimensions, this formula produces a surface whose contours are concentric circles with a Gaussian distribution from the center point. Values from this distribution are used to build a convolution matrix which is applied to the original image. By computing the 2D Gaussian of a 3x3 window, the central pixel's new value is set to a weighted average of pixels surrounding it. The original pixel's value receives the highest weight and neighbouring pixels receive smaller weights as their distance to the original pixel increases. Gaussian function applied to a 2D image through a sliding window over the intensity profile of the image blurs the image by reducing the local sharpness of the pixels. The amount of blurring depends upon the spatial arrangements of the pixels, the size of the kernel and the standard deviation of the Gaussian kernel. Hence successive application of the Gaussian to an image captured as a 2D intensity profile of the surface generates the scale space effect whereby it generates successive images which the human eye perceives while moving away from the object. Gaussian function being exponential in nature does not alter the prime characteristic of the image as the differential / integral / Fourier transformation of the Gaussian results in a Gaussian function itself. Hence it is a potential method to analyze and compare the image in the scale-space without altering its prime characteristics.

6.7.2 Hessian Function

The Hessian function is given in equation 6.18 in the form of a 2D matrix operator. Hessian when applied to a function gives the local curvature of the function. 2D-Hessian manifests itself as a matrix of double differential of the intensity profile of the image. Thus the local undulation of the terrain in the form of intensity profile is captured in the 2D Hessian window of the image as given in equation 6.18. For an image, which is a 2D matrix of intensity values, the Hessian of the image is a square matrix of second-order partial derivatives of the images intensity profile. Given the real-valued function $f(x,y) = I$ the Hessian is computed by a 2D matrix as given below.

$$H(x,y) = \begin{pmatrix} \frac{\partial^2 I}{\partial x \partial x} & \frac{\partial^2 I}{\partial x \partial y} \\ \frac{\partial^2 I}{\partial y \partial x} & \frac{\partial^2 I}{\partial y \partial y} \end{pmatrix} \tag{6.18}$$

6.7.3 Difference of Gaussian

Difference of Gaussian (DoG) is an operation where the pixel-by-pixel difference of the Gaussian convolved gray scale image is obtained. First the gray scale image $I(x, y)$ is smoothened by convolving with the Gaussian kernel with certain standard deviation σ to get

$G(x, y) = G_\sigma(x, y) * I(x, y)$

where

$$G_\sigma = \frac{1}{\sqrt{(2\pi\sigma^2}} exp[-\frac{x^2 + y^2}{2\sigma^2}] \qquad (6.19)$$

Let the Gaussian at two different σ_1 and σ_2 be given by G_{σ_1} and G_{σ_2} respectively. The difference of Gaussian operator smoothes the high gray level intensity of the image profile simulating the scale space as given by following equation.

$$G_{\sigma_1} - G_{\sigma_2} = \frac{1}{\sqrt{(2\pi)}}[\frac{1}{\sigma_1^2}exp^{-\frac{(x^2+y^2)}{2\sigma_1^2}} - \frac{1}{\sigma_2^2}exp^{-\frac{(x^2+y^2)}{2\sigma_2^2}}] \qquad (6.20)$$

DoG has a strong application in the area of computer vision and detection of blobs [44],[43]. Some of the applications of differential geometry in analyzing and visualizing digital images are discussed by Koenderink et al. [28],[29],[30]. Detection of blob refers in computer vision to detecting points and/or regions in the image that differ in properties like brightness or colour compared to the surrounding but have soft boundaries as opposed to crisp boundaries like land and water interface. There are two main classes of blob detectors (1) differential methods based on derivative expressions and (2) methods based on local extrema in the intensity landscape. With the more recent terminology used in the field, these operators can also be referred to as interest point operators, or alternatively interest region operators. There are several motivations for studying and developing blob detectors. One main reason is to provide complementary information about regions, which is not obtained from edge detectors or corner detectors or algebraic method of change detectors. In early work in the area, blob detection was used to obtain regions of interest for further processing. These regions could signal the presence of objects or parts of objects in the image domain with application to object recognition and/or object tracking. In other domains, such as change detection and study of curvature from intensity profile of image, blob descriptors can also be used for peak detection with application to segmentation. Another common use of blob descriptors is as main primitives for texture analysis and texture recognition. In more recent work, blob descriptors have found increasingly popular use as interest points of images with varying curvature that can be classified into flat, elongated or spherical objects.

6.8   Summary

Differential geometry plays a crucial role in computing and analyzing the geometric quantities from gridded data used in GIS. The differential geometric methods Laplacian, Hessian, Gaussian can be modeled as difference equations to compute values from 2D gridded data. Therefore in analyzing images the image is modeled as a 2D array of intensity values often normalized to realize a 3D surface. In this chapter the differential geometry methods such as Laplacian and Hessian are used to compute the gradient and curvature of the intensity surface. The relative value of the eigenvalues computed using the Hessian gives the geometric type of the surface. The interpretation of the eigenvalues of the intensity surface is given for understanding the local geometric property of the image surface.

FIGURE 13.1
Satellite image of Chilka Lake in the state of Odisha in India depicting a land, sea and lake with its vector map draped on it

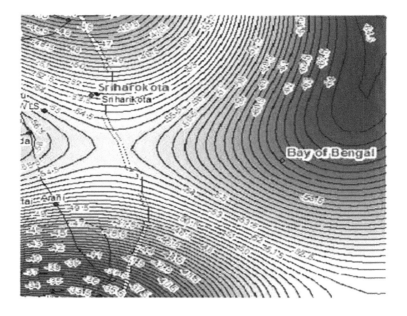

FIGURE 13.2
A contour map covering a portion of land and sea

FIGURE 13.3
Topobathymetry surface with vector data of topography and S-57 bathymetry data of sea

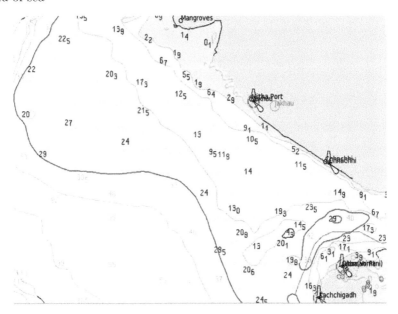

FIGURE 13.4
Topobathymetry surface depicting the sea contours and sounding measures of the sea depth in fathoms

FIGURE 13.5
An instance of a flythrough visualization of a DEM draped with raster map

FIGURE 13.6
3D perspective visualization of an undulated terrain with sun shaded relief
map draped on it

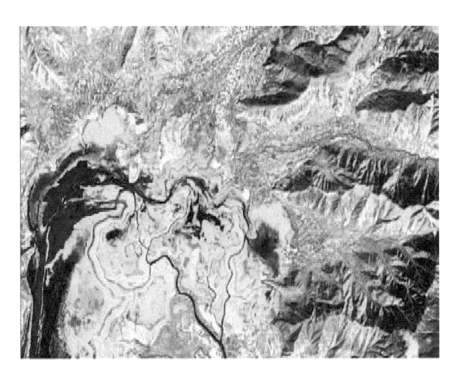

FIGURE 13.7
Colour-coded satellite image of an undulated terrain surface depicting relief

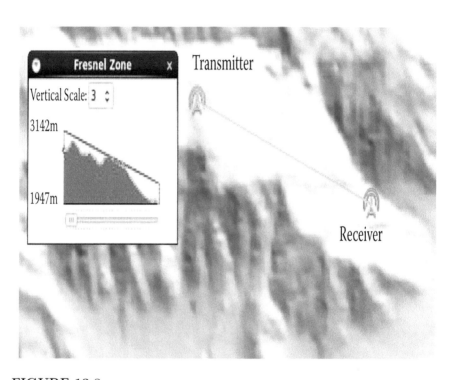

FIGURE 13.8
Computation of communication line of sight between transmitter and receiver
with the corresponding terrain profile along the LOS

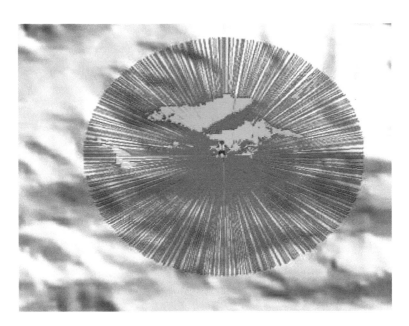

FIGURE 13.9
Computation of line-of-sight fan 360 degrees around the observer

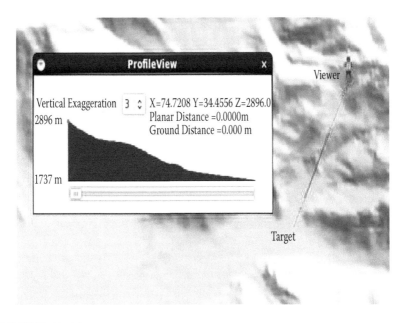

FIGURE 13.10
Line of sight between observer and the target the visible portion is depicted
in green and invisible in red

7

Computational Geometry and Its Application to GIS

Computational geometry is defined broadly as the design and analysis of algorithms for solving problems involving geometric objects. Use of the term 'computational geometry' and its meaning differs in different application contexts. Most researchers in computer science interpret the subject as design and optimization of algorithms involving geometric problems. The term 'algorithm design' carries the connotation of discrete algorithms as opposed to the algorithms used in numerical analysis. The numerical analysis problems are used for solving computational problems in continuous domains. Computational geometry is useful in solving problems of discrete combinatorial geometry rather than continuous geometry. This chapter describes the computational geometric algorithms in the context of their applications in GIS and how these algorithms are used to process spatial data which are the primary inputs of GIS.

Computational geometry emerged as a field of research three decades ago. Since this area has been identified, a number of core geometric problems have emerged. A number of computational geometric algorithms have been designed to solve these problems. These algorithms are optimized for their computation time and memory. Research in computational geometry is progressing to compute large volumes of geometric data and degenerate data. Therefore the robust versions of these algorithms have been devised for processing the degenerate spatial data often encountered by GIS. These algorithms are important for many applications, of great generality and applicability.

7.1   Introduction

Many of the problems that arise in application areas such as computer graphics, computer-aided design and manufacturing, robotics, GIS, computer vision, human-computer interface, astronomy, computational fluid dynamics, molecular biology etc. can be described using discrete geometrical structures. These problems can be solved using a set of algorithms known as computational geometric algorithms. However it should be remembered that not all the geometric applications can be modelled using the discrete geometric structures or solved using computational geometric algorithms. For example, problems in computer vision and computer graphics to some extent are not modelled on the Euclidean plane, but using a matrix of digitized locations of pixels. Problems in computer-aided manufacturing often involve curved surfaces rather than polyhedral surfaces. Problems in fluid dynamics are generally of a continuous nature defined by differential equations. Thus, computational geometry as it is commonly defined is not quite broad enough to address all problems in all these areas. However, the field is broad enough such that virtually all of these application areas can use some of the algorithms in computational geometry. A good introduction to computational geometry along with the algorithms and data structures has been compiled by Preparata and Shamos in [49].

The algorithms in computational geometry discussed in this chapter along with their applications to GIS in the context of processing spatial data pertaining to land, sea and air are listed below.

1. Algorithms to determine line-line intersection.

2. Algorithms to find whether a point lies inside a triangle, polygon, circle, or sphere.

3. Algorithms for computing convex hull.

4. Computing triangulation of a simple polygon in 2D.

5. Computing Delaunay triangulation of a set of points in a plane.

6. Computing the Voronoi tessellation of a set of points in a plane.

These sets of algorithms are often called Computational Geometric Algorithmic Library (CGAL). The input to CGAL algorithms are typically a finite collection of geometric elements such as points(locations associated with place names in the map), lines or line segments (communication lines such as roads, rails, power transmission lines etc.), polygons (coverage of states within the geographic boundary of a country, water bodies), polyhedrons, circles, spheres in the Euclidean space.

7.2   Definitions

7.2.1   Triangulation and Partitioning

Triangulation is a generic method for subdividing a complex domain into a disjoint collection of 'simple' objects. A triangle is the simplest region into which one can decompose a plane and this process is known as triangulation. The higher dimensional generalization of a triangle in 3D is a tetrahedron. Therefore a 3D bounded volume can be decomposed to tetrahedrals. Triangulation or domain decomposition or tessellation is typically a first step performed in number of algorithms. These triangles are then subjected to computational operations through an iterative process to compute the objective and analyze the overall domain.

Triangulation 'T' is tessellation of a polygonal region of the plane into non-overlapping, continuous triangles T_i such that, their intersection is empty, or it is coincident with a vertex, or an edge of both triangles.

$$T = \sum_{i=1}^{|T_i|} T_i \tag{7.1}$$

where, $|t_i|$ is the number of triangles in the domain. Hence the domain of T is the plane curved by its triangles. In addition to the above definition, if the triangulation of the domain is such that, the circumcircle of each triangles does not contains any other points of the domain then it is called Delaunay Triangulation (DT). This is known as the empty circumcircle property of DT.

7.2.2   Convex Hull

Perhaps the first problem in the field of computational geometry is the problem of computing convex hulls, that is, the smallest convex shape that surrounds a given set of objects. In other words, it suffices to say that the convex hull can be imagined as a stretched elastic membrane surrounding the objects which snap tightly around the objects. It is an interesting problem both because of its applications as an initial step towards solving other algorithms, and the number of interesting algorithmic approaches that have been devised to solve this problem.

Mathematically Convex Hull (CH) can be defined through the set theoretic operations as follows.

Let S be a set of discrete objects in 2D plane. The set S is convex if

$$X \epsilon S \wedge Y \epsilon S \Leftrightarrow \bar{XY} \epsilon S \tag{7.2}$$

Generally segment $\bar{xy}$ is the set of points x, y of the form $\alpha x + \beta y = 1$ and $\alpha \geq 0$ and $\beta \geq 0$.

7.2.3   Voronoi Diagram and Delaunay Triangulation

Given a collection of points in space, perhaps the most important geometric data structures for describing the relationships among these points and the relationship of any points in space to these points are very well addressed by structures of the Voronoi diagram and DT. A number of important problems such as the notion of 'proximity' and 'line of sight' can be solved using these two structures. These structures possess a number of beautiful mathematical properties that distinguish them as important geometric structures. Informally, the Voronoi diagram of a finite set of points in the plane is a subdivision of the plane into disjoint regions, one for each point. The Voronoi region of a point consists of the points in the plane for which this point is the closest point of the set. The dual graph of the Voronoi diagram is the Delaunay triangulation. Given a set P of M unique random points in an n-dimensional space, Let us define a region D_i such that

$D_i = \{x : |x - p_i| \le |x - P_j|, \forall(i, j)\}$

Then the collection of the subdivisions D_i is defined as the Voronoi tessellation or Dirichilet tessellation of the set of points P that satisfies the constraint

$$D = \sum_{m=1}^{m} D_m \tag{7.3}$$

7.3   Geometric Computational Techniques

The most important aspect of solving computational geometric problems is learning the design techniques needed in the creation of algorithms. Some of the standard algorithm design techniques pursued in solving any problems in computer science are divide-and-conquer, dynamic programming, greedy technique etc. These techniques work perfectly with alpha numeric data or data where indexing and sorting can be carried out easily. Spatial geometric data often is associated with dimension and randomness. Therefore preprocessing techniques have been developed to bring the spatial geometric data to a representation where ordering and indexing can be applied. Some of these techniques which are often treated as the pre-processing techniques in computational geometry are:

1. **Plane Sweep**
   2-dimensional problems can be converted into a dynamic 1-dimensional problem by sweeping an imaginary line across the place and solving the problem incrementally as the line sweeps across. In general, if you can solve a dynamic version of a (d-1)-dimensional problem efficiently, you can use that to solve a static d-dimensional problem using this technique. A radial sweep line algorithm for construction of Triangular Irregular Network (TIN) has been developed by Mirante et al. [38]. An Implementation of Watsons algorithm for computing 2D Delaunay triangulation is discussed in [46],[47].

2. **Randomized incremental algorithms**

   One of the simplest techniques for the construction of geometric structures is the technique of adding objects to the structure one by one in some order [16]. It turns out that for any data set there may be particularly bad orders in which to insert things (leading to inefficient running times), as well as particularly good orders (leading to efficient running times). It is difficult to know in advance what the proper insertion order of items should be, but it is true for many problems that a random insertion order is efficient with high probability.

3. **Fractional cascading**

   One important technique needed in the design of efficient geometric search problems is that of cascading a sequence of complex decisions up a search tree to generate a longer sequence of simple decisions. This technique has applications in a large number of search problems.

7.4   Triangulation of Simple Polygons

The problem of triangulating polygons can be introduced by way of an example in the field of combinatorial geometry. Combinatorial geometry is the field of mathematics that deals with counting problems in geometry. Combinatorial geometry and computational geometry are closely related, because the analysis and design of efficient geometric algorithms often depends on a knowledge of how many times or how many things can arise in an arbitrary geometric configuration of a given size.

A polygonal curve is a finite sequence of line segments, called edges joined end to end. The endpoints of the edges are vertices. For example, let $v_0, v_1, .., v_n$ denote the set of n+1 vertices, and let $e_0, e_1, .., e_{n-1}$ denote a sequence of n edges, where $e_i = v_i v_{i+1}$. A polygonal curve is closed if the last endpoint equals the first $v_n = v_0$. A polygonal curve is simple if it is not self-intersecting. More precisely this means that each edge e_i does not intersect any other edge, except for the endpoints it shares with its adjacent edges (Figure 7.1).

The famous Jordan curve theorem states that every simple closed plane curve divides the place into two regions i.e. the interior and the exterior regions. Although the theorem seems intuitively obvious, it is quite difficult to prove, and many erroneous proofs were announced before Jordan finally produced a correct proof. A polygon can be defined as the region of the plane bounded by a simple, closed polygonal curve. The term simple polygon is also often used to emphasize the simplicity of the polygonal curve.

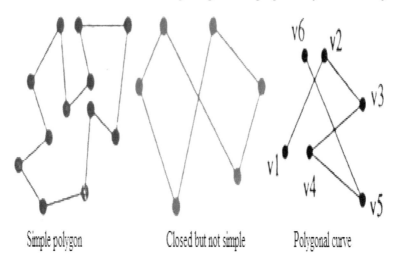

FIGURE 7.1
Polygonal curves

Let us denote the interior of the polygon, as int(P), as an open set which does not contain the boundary. When discussing a polygon P, sometimes it is the interior region of the polygon, that is of interest. Therefore unless explicitly mentioned a polygon means unambiguously the int(P).

O'Rourke makes the rather standard convention that when dealing with polygons, the edges are directed in counter clockwise order about the boundary. Thus the interior of the polygon int(P) is locally to the left of the directed boundary. Such a listing of the edges is called a boundary traversal.

7.4.1   Theory of Polygon Triangulation

Before getting to discussion of algorithms for polygon triangulation, it is pertinent to establish some basic facts about polygons and triangulations. These facts may seem obvious but one must look at them carefully when making geometric arguments. It is quite easy to draw pictures so that a fact appears to be true but in fact the fact is false. For example, many of the facts described here do not hold in 3D space.

Lemma: *Every polygon contains at least one strictly convex vertex (a vertex whose interior angle is strictly less than π).*

Proof: Consider the lowest vertex v in P (i.e. having the minimum y-coordinated). If there are more that one such vertices consider the rightmost. Consider a horizontal line L passing through v. Clearly the vertices adjacent

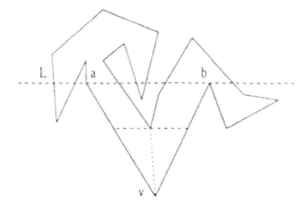

FIGURE 7.2
Existence of a diagonal

to v lie on or above L, and the vertex immediately following v must lie strictly above L. It follows that the interior angle at v is less that π.

Lemma: *Every polygon of $n \geq 4$ vertices has at least one diagonal.*

Proof: Let v be a strictly convex vertex (one exists). Let a and b be the vertices adjacent to v. If ab is a diagonal then we are done. If not, because $n \geq 4$, the closed triangle avb contains at least one vertex of P. Let L be the line through ab. Imagine for concreteness that L is horizontal. See Figure 7.2

Move a line parallel to L through v upwards, and let x be the first vertex hit by this line within the triangle. The portion of the triangle swept by the line is empty of the boundary of P, and so the segment xv is a diagonal.

The above lemma does NOT hold for polyhedral in 3D or higher dimensions.

Lemma: *Every polygon with n vertices can be partitioned into triangles by the addition of (zero or more) diagonals.*

Proof: The proof is by induction on n. If $n = 3$ then the polygon is a triangle and we are done. If not, then by the previous lemma we can find a diagonal. The addition of this diagonal partitions the polygons, each with fewer vertices than the original . By the induction hypothesis, we can partition these into triangles.

Lemma: *The number of diagonals in the triangulation of a n vertex polygon is n-3. The number of triangles is n-2.*

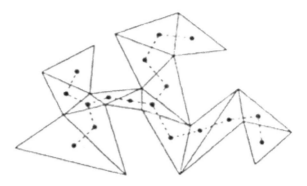

FIGURE 7.3
Dual graph triangulation

Proof: Because the addition of each diagonal breaks the polygon into two, one with k vertices ($k \geq 3$) and the other with $nk + 2$ vertices (the extra 2 are because the two vertices on the diagonal are shared by both parts), the following recurrence relation holds.

$$D(3) = 0$$
$$D(n) = 1 + D(k) + D(n \ k +2) \quad \text{for } 3 \leq k \leq n - 1$$

and

$$T(3) = 1$$
$$T(n) = T(k) + T(n \ k +2) \quad \text{for } 3 \leq k \leq n - 1$$

These can be solved easily by induction on n, $D(n) = n3$ and $T(n) = n-2$. Note that the specific value of k is irrelevant; the argument works for any k in the specified range.

An immediate corollary to the above proof is that the sum of internal angles is $(n - 2)\pi$, because each triangle contributes π to the sum, and the sum of triangle angles is the sum of interior angles.

7.4.2   Dual Tree

A graph can be created in which each vertex of the graph is a triangle of the triangulation, and two vertices in this graph are adjacent if and only if the two triangles share a common diagonal. Such a graph is called the dual graph of the triangulation. An important property of the dual graph is the following.

Lemma: *The dual of a triangulation of a simple polygon is a tree (i.e. a connected, acyclic graph) of degree at most 3.*

Proof: The fact that the degree is at most 3 is a simple consequence of the fact that a triangle can have at most 3 diagonals. It is easy to see that the graph is connected, because the interior of the polygon is connected. To see that there are no cycles, imagine to the contrary that there was a cycle. From this cycle in the dual graph, one can construct a closed path in the interior of the polygon. Either side of this path contains at least one vertex of the polygon. Since each vertex of the polygon is on the boundary of the polygon, this path separates the exterior of the polygon into two disconnected regions. However, by the Jordan curve theorem, the exterior of the polygon is a connected region.

Define an ear to be a triangle of the triangulation that has two edges on the boundary of the polygon, and the other edge is a diagonal. An interesting (and important) fact about polygon triangulations is the following:

Lemma: *Every triangulation of a polygon with $n \geq 4$ vertices has at least 2 ears.*

Proof: An ear is represented as a leaf in the dual tree. Every polygon with at least 4 vertices has at least 2 triangles, and hence at least 2 nodes in its dual tree. It is well known that every (free) tree of at least 2 nodes has at least 2 leaves.

We cannot give the proof that the triangulation of every polygon can be 3 coloured. The proof is by induction. If $n = 3$ then this easy. If $n \geq 4$, then cut off an ear of the triangulation. Inductively colour the remaining triangulation graph. Now restore the ear. The ear shares two vertices in common with the coloured polygon, and so these clolours are fixed. We have exactly no choice for the remaining vertex. (Note that this not only implies that the triangulation graph is 3 colourable, but up to a permutation of the colours, the colouring is unique.)

7.4.3 Polygon Triangulation

The discussion of the implementation of a very simple (but asymptotically inefficient) algorithm for polygon triangulation is continued in [39], [40].

7.4.3.1 Order Type

The fundamental primitive operation upon which most of the theory of 1-dimensional sorting and searching is based is the notion of ordering, namely that numbers are drawn from a totally ordered domain, and this total ordering can be used for organizing data for fast retrieval.

Unfortunately, there does not seem to be a corresponding natural notion of ordering for 2 and higher-dimensional data. However, interestingly there is still a notion of order type.

Given two (1-dimensional) numbers, a and b, there are 3 possible relationships, or order types, that can hold between them:

$$a<b, a=b, a>b$$

This relation can be described as the sign of the quantity, ab. More generally, in dimension 2, we can define the order type of three points, a, b, and c as:

1. a, b, and c form a clockwise triple.

2. a, b, and c are collinear.

3. a, b, c form a counter clockwise triple.

Similarly, for four points in dimension 3, we can define the order type of these points as: they form a left-handed screw, they are coplanar, or they form a right-handed screw. Interestingly, all of these order-relations fall out from a single determinant of the points represented. These can be illustrated for pairs a and b in 1-space, triples a, b, and c in the plane, and quadruples a, b, c and d in 3 space. The coordinates of a point a are denoted $(a_0, a_1, a_2, ..., a_{d-1})$.

$$Ord(a, b) = \begin{vmatrix} a_0 & 1 \\ b_0 & 1 \end{vmatrix}$$

$$ord(a, b, c) = \begin{vmatrix} a_0 & a_1 1 \\ b_0 & b_1 1 \\ c_0 & c_1 1 \end{vmatrix}$$

$$ord(a, b, c, d) = \begin{vmatrix} a_0 & a_1 a_2 1 \\ b_0 & b_1 b_2 1 \\ c_0 & c_1 c_2 1 \\ d_0 & d_1 d_2 1 \end{vmatrix}$$

It is well known that the value of Ord(a,b,c) in the plane is the signed area of the parallelogram defined by these vectors, or twice the area of the triangle defined by these points. The area is signed so that if the points are given in counter clockwise order the area is positive, and if given in clockwise order the area is negative. In general, it gives d times the signed volume of the simplex defined by the points. An interesting consequence of this is the following formula.

Theorem: *Area of a polygon Given a simple polygon P with vertices* $v_0, v_1, ., v_{n-1}$, *where* $v_i = (x_i, y_i)$ *the area of P is given by:* $2A(P) = \sum_{i=0}^{n-1}(x_i y_i + 1 - y_i x_i + 1)$

Proof: The proof comes about by computing the sums of the signed area of the triangles defined by any point in the plane (e.g. the origin) and the edges of the polygon, and expanding all the determinants given above, and then simplifying.

This theorem can also be generalized for computing volumes, where the term in the summation is over the determinants of the simplicial faces (e.g. triangles in 3-space) of the boundary of the polyhedron.

Note that in low dimensional space (2 and 3) these determinants can be computed by simply expanding their definitions. In higher dimensions it is usually better to compute determinants by converting it into upper triangular form by, say, Gauss elimination method, and then computing the product of the diagonal elements.

7.4.4   Line Segment Intersection

In the naive triangulation algorithm, an important geometric primitive which needs to be solved is computing whether two line segments intersect one another. This is used to determine whether the segment joining two endpoints v_i and v_j is a diagonal. The problem is, given two line segments, represented by their endpoints, (a, b) and (c, d), to determine whether these segments intersect one another. The straight forward way to solve this problem is to

1. Determine the line equations on which the segments lie. (Beware: If either of the lines are vertical, the standard slope/intercept representation will be undefined.)

2. Determine the (x,y) coordinates at which these two lines intersect. (Beware: If the lines are parallel, or if they coincide, these must be treated as special cases. Observe that even if the input points are integers, this point will generally be a non-integer [rational] value.)

3. Determine whether this point occurs within both of the line segments.

In spite of the obvious importance of this simple primitive, the above code is rather complex and contains a number of special cases which require careful coding. A number of the different types of segments and special cases to consider are shown below.

A methods which uses only integer arithmetic (assuming the input coordinates are integers) and avoids any of these problems is proposed. A test is built up by combining primitive operations, starting with the orientation test. Because the orientation test returns twice the signed area of the triangle, O'Rourke [40] calls it Area2.

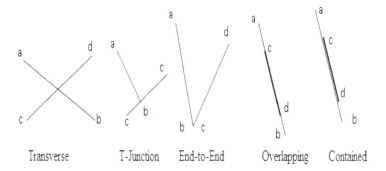

FIGURE 7.4
Types of line segment intersections

Int Area2(a, b, c) { return a[0]*b[1] a[1]*b[0]+ a[a]*c[0] [0]*c[1]+ b[0]*c[1] b[1]*c[0]; }
The first primitive is a test whether a point *c* lies to the left of the directed line segment $\bar{ab}$.
bool Left(a, b, c)
{
return Area2(a, b, c) >0; // is c left of ab?
}
bool Lefton(a, b, c)
{
return Area2(a, b,c) >=0; // is c left of or on ab?
}
bool Collinear(a, b, c)
{
return Area(a, b, c) == 0; // is c on ab?
}

To determine whether two line segments intersect, one can begin by distinguishing two special cases, one is where the segments intersect transversely (they intersect in a single point in their interiors). O'Rourke calls this a proper intersection. This is the normal case, so we test for it first. Notice that if there is a non-proper intersection (collinear lines, T-junction, or end-to-end), then at least 3 points must be collinear. Otherwise, it suffices to test that the points *c* and *d* lie on opposite sides of the line segment $\bar{ab}$. From this we get the following code for proper intersection.

bool IntersectProp(a, b, c, d)
{

```
If((Collinear(a,b,c) ||Collinear(a, b, d) ||Collinear(c, d, a) ||Collinear(c, d, b))
return False;
else
return Xor((Left(a, b, c), Left(a, b, d)) && Xor(Left(c, d, a), Left(c, d, b));
}
```

The function Xor is the eXclusive-or operator, which returns 'True' if one and only one of the arguments is True. For improper intersections, it is needed to test whether one of the endpoints of 'cd' lies within the segment $a\bar{b}$, or vice versa. This can be done by a betweenness test, which determines whether a point 'c' is collinear and lying on the segment $a\bar{b}$ (including the segment's endpoints). The overall intersection test just tests for proper and improper intersections.

```
bool Between(a, b, c)
{
If( ! Collinear(a, b, c)) return False; // not collinear
If((a[X] ! = b[X])
return ((a[x] <= c[x] && c[x] <= b[x])) ||((a[x] >= c[x] && (c[x] >= b[x]));
else
return(( a[y] <= c[y] (c[y] <= b[y])) ((a[y] >= c[y] && ( c[y] >= b[y]));
}
```

```
bool Intersect(a, b, c, d)
{
If(IntersectProp(a, b, c d)) return True;
else if(Between(a, b, c) ||Between(a, b, d) ||Between(c, d, a) ||Between(c, d, b)) return True;
return False;
}
```

7.4.5 Finding Diagonals in a Polygon

One can use the code above to test whether a line segment $v_i v_j$ in a polygon is a diagonal Figure. 7.5. The test consists of two parts.

1. Test whether $v_i v_j$ intersects any edge $v_k v(k+1)$ along the boundary of the polygon. (Note that indices are taken modulo n.) The edges that are adjacent to v_i and v_j should not be included among these edges.
2. Test whether the segment $v_i v_j$ lies interior to the polygon locally

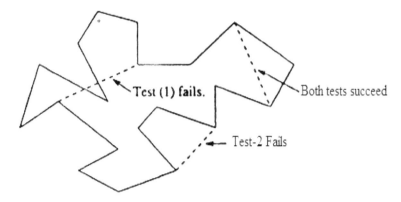

FIGURE 7.5
Diagonal test in a polygon

> in the vicinity of v_i and v_j. This is needed to discard segments that
> lie entirely on the exterior of the polygon.

The second test can be performed using the same left predicate. Details
are in O'Rourke [40]. Observe that intersection testing runs in O(1) time,
and since we repeat this against every edge of the polygon the whole test is
O(n) in time, where n is the number of vertices in the polygon. A detailed
discussion of implementing computational geometric algorithms using c and
C++ computer language has been discussed by Joseph O'Rourke in his book
Computational Geometry Using C [40].

7.4.6   Naive Triangulation Algorithm

A simple but not very inefficient triangulation algorithm for triangulating a
polygon can be designed using the diagonal finding test. Test for each potential
diagonal, v_iv_j, using the diagonal test. Each such test takes O(n) time. When
you find a diagonal, remove it. Split the polygon into two sub-polygonals,
times O(n) for each diagonal test, followed by two recursive calls. Since there
are a total of (n - 3) diagonals to be found, one can argue that the worst-case
running time for this algorithm is O(n^4).

An improvement to the above method can be done by observing that the
test needs to be performed for ear-diagonals. We consider each ear-diagonal,
v_{i-1}, v_{i+1}, and apply our diagonal test. Each such test takes O(n) time. Since
there are only O(n) potential ears, and hence O(n) potential ear-diagonals,
this takes O(n^2) time. We still need to apply this O(n) times to find diagonals,
leading to an O(n^3) algorithm.

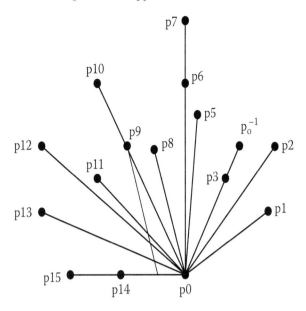

FIGURE 7.6
Graham's scan

7.5   Convex Hulls in Two Dimensions

In this section three algorithms, the Graham's scan algorithm, the divide and conquer algorithm and the QuickHull algorithm, are discussed. These algorithms compute the convex hull of a set of random points given in the 2D plane.

7.5.1   Graham's Scan:

The first $O(n \log n)$ worst-case algorithm for computing the convex hull for a set of points in 2D was developed by Graham. The idea is to first topologically sort the points, and then use this ordering to organize the construction of the hull. How do we topologically sort the points? A good method is to sort cyclically around some point that lies either inside or on the convex hull. A convenient choice is lowest point (and rightmost if there are ties). Call this point p_0. Compute the angle formed between a horizontal ray in the $+x$ direction from this point and each point of the set. Sort the points by these angles. (If there are ties, we place points closer to p_0 earlier in the sorted order.)

We will walk through the points in sorted order, keeping track of the hull formed by the first point seen so far. These points will be stored in a stack.

(O'Rourke suggests using a linked list to represent this stack.) The essential issue at each step is how to process the next point, p_i. Let p_t and p_{t-1} denote the vertices that currently at the top and next to top positions on the stack. If p_i lies strictly to the left of the directed segment $p_{t-1}p_t$ (which we can determine by an orientation test) then we add p_i to the stack. Otherwise, we know that p_t cannot be on the convex hull. Why not? (Because the point p_t lies within the triangle $p_o p_{t-1} p_i$, and hence cannot be on the hull.) We pop p_t off the stack. We repeat the popping until p_i is strictly to the left of the previous hull edge. Graham's scan algorithm has been discussed by Preparata et al. in [49].

7.5.1.1   Steps of Graham's Scan

1. Find the rightmost lowest point p[0];
2. Sort points angularly about p[0], store in p[1],p[2],.. ,p[n 1];
3. Stack S = Φ. Push n - 1. Push 0. i = 1;
4. While i <n do:

    •if p[i] is strictly left of (p[S[(top 1]],p[S[t]]) then push i and i = i+ 1;
    •otherwise pop S;

5. Pop S (since p[n 1] was pushed twice).
6. Output the contents of S.

To analyze the running time of Graham's scan, observe that we spend $O(n \log n)$ initially in sorting the points. Each subsequent iteration of the loop takes $O(1)$ time to either (1) push a new point on the stack of hull vertices, or (2) pop a point off this stack. Observe that each point in the data set can be pushed at most once, and popped at most once. Therefore, the total number of pushes and pops is at most 2n. Since each can be processed in $O(1)$ time, the overall time is $O(n \log n)$ for sorting plus $O(n)$ time for hull computation.

The implementation of Graham's scan is given in O'Rourke. In its implementation, one thing which is an important primitive is that of computing angles while sorting the points. The standard method for doing this is to compute the vectors $r_i = p_i - p_o$ for $1 \leq i < n$. The angle we desire is essentially the slope of the vector r_i, which can be computed using a inbuilt function in standard mathlib procedure atan2(). An improvement and better method is not to compute angles at all, but work with slopes. It can be observed that to sort the vectors r_i according to slope the basic comparison which is needed is whether

$$\frac{r_i[X]}{r_i[Y]} < \frac{r_j[X]}{r_j[Y]} \tag{7.4}$$

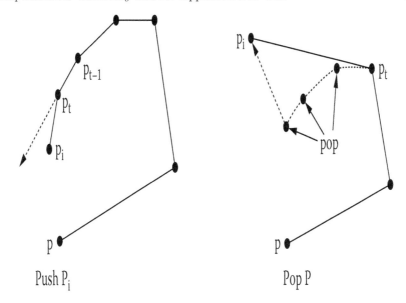

FIGURE 7.7
Push and pop operation

Recall that because of the choice of p_0, all the y-coordinates of these vectors are non-negative. Thus we can convert this into an equivalent integer condition:

$$r_i[X] * r_j[Y] < r_j[X] * r_i[Y] \qquad (7.5)$$

A further improvement to this method is to use the orientation test already discussed in the previous section. To compare p_1 with p_2 in the ordering, it suffices to test the orientation of the triple (p_0, p_1, p_2). If positive then $p_1 < p_2$, if negative then $p_2 < p_1$, and if zero, we need to compute the vectors r_1 and r_2 and compare their (squared) lengths. This has the advantage of using the already constructed tools.

7.6   Divide and Conquer Algorithm

Another O(n logn) algorithm for computation of convex hull is based on the divide and conquer strategy. This algorithm can be viewed as a generalization of the merge sort algorithm. The outline of the algorithm is given below.

7.6.1   Divide and Conquer Convex Hull

1. Sort the points by x-coordinate.

2. Divide the points into two sets A and B, where A consists of the left [n/2] points and B consists of the right [n/2] points.

3. Recursively compute $H_A = \mathrm{conv}(A)$ and $H_B = \mathrm{conv}(B)$. (Note: The sorting step does not need to be repeated here.)

4. Merge the two hulls into a common convex hull, H and output this hull.

The recursion bottoms out when the current set consists of three or fewer points, in which case it is trivial to compute the hull in O(1) time. Clearly the initial sorting takes O(n logn) time. For the rest of the algorithm, the running time is given by the recurrence relation,

$$T(n) = f(n) + 2T\left(\frac{n}{2}\right) \tag{7.6}$$

It is known that this recurrence solves to T(n) ϵ O(nlogn) if f(n) ϵ O(n). Thus, we need to figure out how to merge two convex hulls of size n/s in time O(n). One thing that simplifies the algorithm is the knowledge that the hulls are separated from each other by a vertical line (assuming no duplicate x-coordinates). The merging process boils down to computing two tangent lines, an common upper tangent and a common lower tangent to the two hulls and then discarding the portions of the hulls lying between these tangents. This process is depicted in Figure 7.8.

How are these tangents computed? Let's concentrate on the lower tangent. The upper tangent is similar. The algorithm operates by a simple walking procedure. We initialize a to be the rightmost point of H_A and b is the leftmost point of H_B. (These can be found in linear time.) Lower tangency is a condition that can be tested locally by an orientation test of the two vertices and neighbouring vertices on the hull. (This is a simple exercise.) We iterate the following two loops, which march a and b down, until they reach the point's lower tangency.

7.6.1.1   Lower Tangent

1. Let a be the rightmost point of H_A

2. Let b be the leftmost point of H_B

3. While ab is not a lower tangent for H_A and H_B do:

   While ab is not a lower tangent to H_A do a = a + 1; (move a clockwise)

   While ab is not a lower tangent to H_B do b = b +1 ; (move b counter clockwise)

4. Return ab

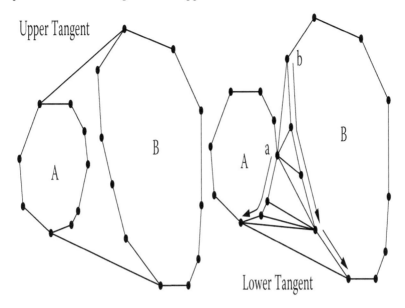

FIGURE 7.8
Computing the lower tangent

7.6.2 Quick Hull

One technique for constructing geometric algorithms for 2-dimensional problems is to attempt to generalize some 1-dimensional algorithm. Here is one method for generalizing QuickSort, called QuickHull. Like QuickSort, this algorithm runs in O(n logn) time for favorable inputs and $0(n^2)$ time for unfavorable inputs. However, unlike QuickSort, there is no obvious way to convert it into a randomized algorithm with 0(n logn) expected running time. However, like QuickSort, this algorithm tends to perform very well in practice, because the worst-case scenarios tend to be rare.

The intuition is that most of the points lie within the hull, rather than on its boundary, so think of a method that discards interior points as quickly as possible. QuickHull begins by computing the points with the maximum and minimum, x- and y-coordinates. Clearly these points must be on the hull. Horizontal and vertical lines passing through these points are support lines for the hull, and so define a bounding rectangle, within which the hull is contained (Figure 7.9). Furthermore, the convex quadrilateral defined by these four points lies within the convex hull, so the points lying within this quadrilateral can be eliminated from further consideration. All of this can be done in 0(n) time.

To continue the algorithm, we classify the remaining points into the 1 corner triangles that remain. In general, as this algorithm executes, we will have an inner convex polygon, and associated with each edge we have a set of

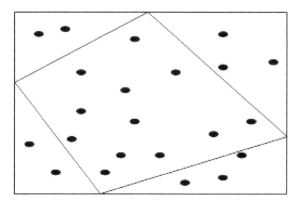

FIGURE 7.9
QuickHulls initial quadrilateral

points that lie outside of that edge. (More formally, these points are witness to the fact that this edge is not on the convex hull, because they lie outside the half-plane defined by this edge.) When this set of points is empty, the edge is a final edge of the hull. Consider some edge 'ab'. Assume that the points that lie outside of this hull edge have been place in a bucket that is associated with 'ab'. The task is to find a point 'c' among these points that lies on the hull, discard the points in the triangle 'abc', and split the remaining points into two subsets, those that lie outside 'ac' and those than lie outside of 'cb'. This process is depicted in the Figure 7.10 (a) and (b).

How should 'c' be selected? There are a number of possible selection criteria that one might think of. The suggested method is that c be the point that maximizes the perpendicular distance from the line ab. (Another possible choice might be the point that maximizes the angle cba or cab. It turns out that these are poor choices because they do not produce even splits of the remaining points.) We replace the edge ab with the two edges ac and cb, and classify the points as lying in one of 3 groups: those that lie in the triangle abc, which are discarded, those that lie outside of ac, and those that lie outside of cb. We put these points in buckets for these edges, and recurse. (We claim that it is not hard to classify each point p, by computing the orientations of the triples 'acp' and 'cbp'.)

The running time of QuickHull algorithm along with QuickSort, depends on how evenly the points are split at each stage. If we let T(n) denote the running time on the algorithm, where the n is the number of points that remain in the current bucket, then the time is given by the recurrence:

$$T(0) = 1$$

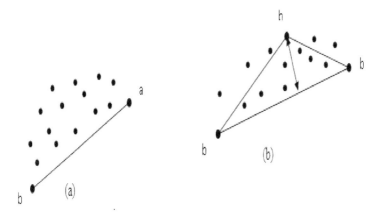

FIGURE 7.10
QuickHull elimination procedure

$$T(n) = cn + T(n_1) + T(n_2)$$

where, n_1 and n_2 are the number of points remaining in the two buckets. It should be a familiar fact from the QuickSort analysis that this running time will be good as long as $\max(n_1, n_2)$ is not too close to n.

7.7 Voronoi Diagrams

A Voronoi diagram (like convex hull) is one of the most important structures in computational geometry. A Voronoi diagram records information about the spatial relationship among the objects such as which spatial object is close to what other spatial object or objects.

Let $P = p_1, p_2, ..., p_n$ be a collection of points in the plane. For a given point 'q' in the plane, the nearest neighbour of 'q' is the point in 'P' whose distance from 'q' is minimum. (In general there can be more than one nearest neighbour if points are equidistant from q.) Define $V(p_i)$, the Voronoi polygon for p_i, to be the set of points in the plane for which p_i is the nearest neighbour of that point. If we let $| pq |$ denote the distance from 'p' to 'q', then we can state this as:

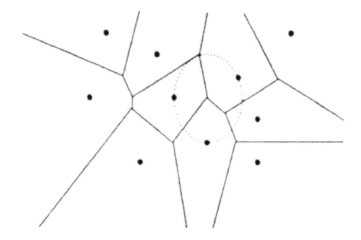

FIGURE 7.11
Voronoi diagram

$$V(p_i) = \{q \mid p_i - q \mid \leq \mid p_j - q \mid \forall j \neq i\} \tag{7.7}$$

The Voronoi polygons subdivide the plane into a collection of regions. (The term polygon is used in a more general sense than defined earlier, because these polygons may be unbounded, stretching to infinity.) The interiors of these regions are pair-wise disjoint, and the boundaries correspond to points in the plane whose nearest neighbour is not unique. The union of the boundaries of the Voronoi polygons is called the Voronoi diagram of P, denoted by VD(P). An example of the Voronoi diagram is depicted in Figure 7.11.

A detail implementation of the Voronoi diagram or Dirichlet tessellation is discussed by Bowyer [5], Green [23] and Fortune [17]. A survey on the Voronoi diagram as a geometric data structure has been carried out by Aurenhammer [3].

7.7.1 Properties of Voronoi Diagrams

Some theoretical observations about the Voronoi diagram are as follows.

1. **Half Plane Formulation:** Recall from high school geometry that the set of points that are equidistant between two points is just the perpendicular bisector. In general, the set of points that are closer to point p_i, than p_j is the half-plane lying to one side of this bisector, $H(p_i, p_j)$. It is easy to see that, $V(p_i)$ can be defined as the intersection of these half-planes.

$$V(p_i) = \bigcap_{j \neq i} H(p_i, p_j) \tag{7.8}$$

2. **Convex:** It is well known that the intersection of two convex sets is convex. Therefore, since half-planes are convex sets, $V(p_i)$ is a convex set, and hence a convex polygon (but possibly unbounded).

3. **Voronoi Vertices:** The vertex at which three Voronoi cells $V(p_i)$, $V(p_j)$ and $V(p_k)$ intersect must be equidistant from all three. Thus it is the center of the circle passing through these points. This circle can contain no other points (since by definition, these are the three closest points to this vertex).

4. **Convex Hull:** A cell of the Voronoi diagram is unbounded if and only if the corresponding site lies on the convex hull. (Observe that a point is on the convex hull if and only if it is the closest point from some point at infinity.)

7.8 Delaunay Triangulation

Observe in Figure 7.12 that if points are in general position, and in particular, if no four points are co-circular, then each Voronoi vertex is incident to exactly three edges of the Voronoi diagram. Since the Voronoi diagram is just a planar graph, we can consider its dual graph. In particular, we connect two points of P in this dual graph if and only if they share an edge in the Voronoi diagram. If the points are in general position, the faces of the resulting dual graph (except for the exterior face) will be triangles (because the Voronoi vertices have degree 3).

An efficient algorithm for generation of mesh using Delaunay triangulation has been developed by Watson and reported in [51]. An implementation of Watson's algorithm for generation of 2D Delaunay triangulation has been discussed in [38].

7.8.1 Properties of Delaunay Triangulation

Delaunay triangulations have a number of interesting properties, that are consequences of the structure of the Voronoi diagram.

1. **Convex Hull:** The exterior face of the Delaunay triangulation is the convex hull of the point set.

2. **Circumcircle Property:** The circumcircle of any triangle in the

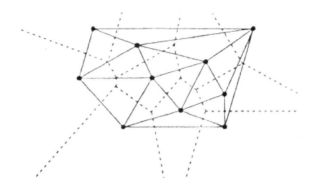

FIGURE 7.12
Delaunay triangulation

Delaunay triangulation is empty (contains no other points of the domain).

3. **Empty Circle Property:** Two points p_i and p_j are connected by an edge in the Delaunay triangulation, if there is an empty circle passing through p_i and p_j. (One direction of the proof is trivial from the circumcircle property. In general, if there is an empty circumcircle passing through p_i and p_j, then the center c of this circle is a point on the edge of the Voronoi diagram between p_i and p_j, because c is equidistant from each of these points and there is no closer point.)

4. **Max-Min Angle Criterion:** Among all triangulations of the point set P, the Delaunay triangulation maximizes the minimum angle in the triangulation. (This property of the Delaunay triangulation is valid in 2D only.)

5. **MST Property:** The minimum spanning tree of a set of points in the plane is a subgraph of the Delaunay triangulation.

This observation is used to develop a good algorithm to compute MST of a set of points in the 2D plane. First compute the Delaunay triangulation of the point set which can be computed in O($n log n$) time. The ortho centers of the Delaunay triangles are computed. These points are further taken as inpur to Kruskal's algorithm for computing the MST. This process takes O($n log n$) time to compute the minimum spanning tree of this sparse graph by joining the ortho centers of the delaynay triangles by shortest medial axis.

Further Delaunay triangulation can be used to compute and generate three different types of graph structures. The Gabriel graph

(GG) which is defined as follows: two points p_i and p_j are connected by an edge if the circle with diameter $p_i p_j$ is empty. The relative neighbourhood graph (RNG) which is defined as: two points p_i and p_j are connected by an edge if there is no point which is simultaneously closer to p_i and p_j. The relationship between these graphs is given by the following set relationship:

$$MST \subseteq RNG \subseteq GG \subseteq DT. \qquad (7.9)$$

7.9 Delaunay Triangulation: Randomized Incremental Algorithm

A simple yet a powerful design technique is randomized incremental algorithm. randomized incemental algorithm for computing Delaunay triangulations from a set of points in the 2D plane has many advantages listed below.

1. It is very simple to understand and implement.
2. It is not hard to generalize it to higher dimensions.
3. It has a simple (backwards) analysis.

Another feature of the algorithm is that it can be modified to produce a data structure for performing point location queries. This is the problem of determining which triangle of the final triangulation contains the query point in O($log n$) time. Because triangulations are often used for other purposes, having a point location data structure is a nice additional feature. (We will show later that point location data structures can be built separately.) The algorithm is not hard to adapt for computing Voronoi diagrams as well (but the associated point location algorithm runs somewhat more slowly, in O(log^2 n) time).

As with any randomized incremental algorithm, the idea is to insert points in random order, one at a time, and update the triangulation with each new addition. The issues involved with the analysis will be showing that the number of structural changes in the diagram is not very large. As with other incremental algorithms, we need some way of keeping track of where newly inserted points are to be placed in the diagram. As we did with trapezoidal decomposition, this can be done by bucketing the points according to the triangles they lie in. In this case, we will need to argue that the number of times that a point is reclassified on average is not too large.

7.9.1 Incremental Update

The basic issue in the design of the algorithm is how to update the triangulation when a new point is added. In order to do this, we first investigate the basic properties of a Delaunay triangulation. Recall that a triangle $\triangle abc$ is in the Delaunay triangulation, if and only if the circumcircle of this triangle contains no other point in its interior. (We will make the usual general position assumption that no 4 points are co-circular.) How to test whether a point d lies within the interior of the circumcircle of $\triangle abc$? It turns out that this can be reduced to a computable determinant. The point d lies within the circumcircle defined by $\triangle abc$ if and only if:

$$\text{In}(a,b,c,d) = \det \begin{pmatrix} a_x & a_y & a_x{}^2 + a_y{}^2 & 1 \\ b_x & b_y & b_x{}^2 + b_y{}^2 & 1 \\ c_x & c_y & c_x{}^2 + c_y{}^2 & 1 \\ d_x & d_y & d_x{}^2 + d_y{}^2 & 1 \end{pmatrix} < 0 \qquad (7.10)$$

Assuming that this primitive In(a,b,c,d) is available to us when we add the next point, p_i, the major steps used by the algorithm to convert the current triangulation into the new triangulation is achieved by the following two steps:

1. Adding a point to the middle of a triangle, and creating three new edges

2. Swapping one or more edges of the triangulation to restore the triangulation as Delaunay triangulation.

Both of these operations can be performed in O(1) time, assuming a winged-edge or quad-edge representation of the triangulation (Figure 7.13).

Here is how the algorithm works. We start with an initial triangulation. Guibas, Knuth and Sharir suggest starting with a triangle of three points 'at infinity'. A somewhat more direct approach is to enclose your points within a large bounding rectangle, and add either of the diagonals. This will be a Delaunay triangulation. Either guarantees that all points to be added, will lie within some triangle of the triangulation.

The points are added in random order. When a new point p is added, we find the triangle abc of the current triangulation that contains this point, insert the point in this triangle, and join this point to the three surrounding vertices. This creates three new triangles, pab, pbc, and pca, each of which may or may not satisfy the empty-circle condition. How to test this? For each of the triangles that have been added, we check the vertex of the triangle that lies on the other side of the edge that does not include p. If this vertex fails the incircle test, then we swap the edge (creating two new triangles that are adjacent to p) and repeat the same test with these triangles. An example of the process is depicted in Figure 7.14.

The following is a description of the algorithm (Guibas, Knuth, and Sharir give a nonrecursive version). The current triangulation is kept in a global data structure. The edges in the following algorithm are actually pointers to the quad-edge data structure.

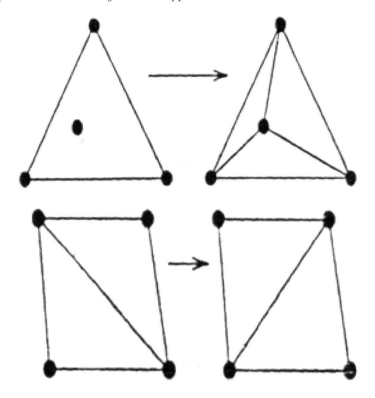

FIGURE 7.13
Basic triangulation changes

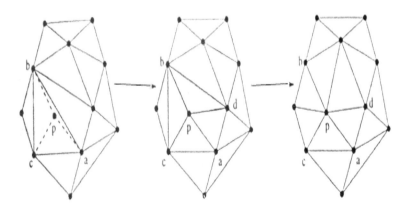

FIGURE 7.14
Point insertion

```
    Insert(p)
{
Find the triangle Δabc containing p;
Insert edges pa, pb, and pc into triangulation;
SwapTest(ab);
SwapTest(bc);
SwapTest(ca);
}
SwapTest(ab)
{
If (ab is an edge on the exterior face) return;
Let d be the vertex to the right of edge ab;
If(in(p. a, b, d))
{
Replace edge ab with pd;
SwapTest(ad);
SwapTest(db);

    }
}
```

As one can observe, the algorithm is very simple. The data structures and the routines that need to be implemented are:

1. The quad-edge or winged-edge data structure operations.

2. The incircle test.

3. Locating the triangle that contains the point 'P' to be inserted to the triangulation.

Task (1) is easy to implement once you have an implementation of the quad-edge or winged-edge data structure. Task (2) is easy. Task (3) can be accomplished by classifying each point according to the triangle that it lies in (point-inside-a-triangle). When an edge is swapped, or when a triangle is split into three triangles through point insertion, the points associated with the affected triangles need to be reclassified. We will discuss an alternative method based on a history search later.

There is only one major issue in establishing the correctness of the algorithm. When we performed empty-circle tests, we only tested:

1. Triangles containing the point p.

2. Only points that lay on the opposite side of an edge of such a triangle.

To establish (1), observe that it suffices to consider only triangles containing p because since p is the only newly added point. Therefore it is the only point that can cause a violation of the empty-circumcircle property.

To establish (2) we argue that if for every point d, which is opposite from p along some edge ab, lies outside the circumcircle of pab, then all these circumcircles are empty. A complete proof takes some effort, but here is a simple justification. What could go wrong? It might be that d lies outside the circumcircle, but there is some other point (e.g. a vertex e of a triangle adjacent to d that lies inside the circumcircle). This is illustrated in Figure 7.14.

7.10   Delaunay Triangulations and Convex Hulls

At first, Delaunay triangulations and convex hulls appear to be quite different structures; one is based on metric properties (distances) and the other on affine properties (collinearity, coplanarity). It can be shown that it is possible to convert the problem of computing a Delaunay triangulation in dimension D to that of computing a convex hull in dimension $(D + 1)$. Thus, there is a remarkable relationship between these two structures.

It can be demonstrated that the Delaunay triangulation in dimension two can be constructed by computing a convex hull in dimension three. This may be hard to visualize, but can be proved through geometrical means. This also can be reasoned by an analogy in one lower dimension of Delaunay triangulations in $1D$ and convex hull in $2D$. The complexities of the structures are not really apparent in this case.

The connection between the two structures is the paraboloid $z = x^2 + y^2$. Observe that this equation defines a surface whose vertical cross sections (constant x or constant y) are parabolas and whose horizontal cross sections (constant z) are circles. For each point in the plane (x, y), the vertical projection of this point, onto this paraboloid is (x, y, x^2, y^2) in 3-space. Given of points S in the plane, let S' denote the projection of every point in S onto this paraboloid consider the lower convex hull of S'. This is the portion of the convex hull of S which is visible to a viewer standing at $z = -\infty$. We claim that if we take the lower convex hull of S', and project back onto the plane, then we get the Delaunay triangulation of S. In particular, let $p, q, r \epsilon S$ and let p, q, r denote the projections of these points onto the paraboloid. Then $p'q'r'$ define a fixed convex hull of S' if and only if Δpqr is a triangle of the Delaunay triangulation of S. The process is illustrated in Figure 7.15.

The question is, why does this work? To find out, we need to establish the connection between the triangles of the Delaunay triangulation and the faces of the convex hull of transferred points. In particular, recall conditions of Delaunay triangulation and convex hull.

- **Delaunay condition:** Three points p, q and r ϵ S form a Delaunay triangle if and only if the circumcircle of these points contains no other point of the domain S.

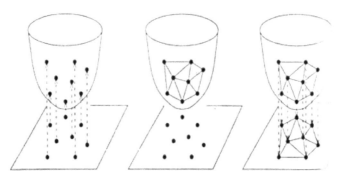

Project on paraboloid   Compute convex hull   Project of hull faces back to plane

FIGURE 7.15
Delaunay triangulations and convex hulls

- **Convex hull condition:** Three points p', q', and r' ϵ S' form a face of the convex hull of S if and only if the plane passing through p', q' and r' has all the points of S' lying to one of its side.

Clearly, the connection we need to establish is between the emptiness of circumcircles in the plane and the emptiness of halfspaces in 3-space. It can be proven by the following claim.

Lemma: *Consider four distinct points p,q,r,s in the plane, and let p', q', r' and s' be their respective projections onto the paraboloid, $z = x^2 + y^2$. The point s lies within the circumcircle of p,q,r if and only if s lies on the lower side of the plane passing through p', q', r'.*

Proof: To prove the lemma, first consider an arbitrary (non-vertical) plane in 3-space, which we assume is tangent to the paraboloid above some point (a, b) in the plane. To determine the equation of this tangent plane, we take derivatives of the equation $z = x^2 + y^2$ with respect to x and y giving

$$\frac{\partial z}{\partial x} = 2x \qquad (7.11)$$

and

$$\frac{\partial z}{\partial y} = 2y \qquad (7.12)$$

At the point $(a, b, a^2 + b^2)$ these evaluate to $2a$ and $2b$. It follows that the plane passing through these point has the form

$$Z = 2ax + 2by + \gamma \qquad (7.13)$$

To solve for γ we know that the plane passes through $(a, b, a^2 + b^2)$ so by solving it gives

$a^2 + b^2 = 2a.a + 2b.b + \gamma$

$$\Rightarrow \gamma = -(a^2 + b^2) \tag{7.14}$$

Thus the plane equation is

$z = 2ax + 2by - (a^2 + b^2)$

If we shift the plane upwards by some positive amount r^2 we get the plane

$z = 2ax + 2by - (a^2 + b^2) + r^2$

How does this plane intersect the paraboloid? Since the paraboloid is defined by $z = x^2 + y^2$ we can eliminate z giving:

$x^2 + y^2 = 2ax + 2by - (a^2 + b^2) + r^2$

which after some simple rearrangements is equal to

$$(x - a)^2 + (y - b)^2 = r^2 \tag{7.15}$$

This is just a circle (Figure 7.16). Thus, we have shown that the intersection of a plane with the paraboloid produces a space curve (which turns out to be an ellipse), which when projected back onto the (x, y)-coordinate plane is a circle centered at (a, b).

Thus it can be concluded that the intersection of an arbitrary lower half-space with the paraboloid, when projected onto the (x, y)-plane is the interior of a circle. Going back to the lemma, when we project the points p, q, r onto the paraboloid, the projected points p, q and r define a plane. Since p, q and r lay at the intersection of the plane and paraboloid, the original points p, q, r lie on the projected circle. Thus this circle is the (unique) circumcircle passing through these p, q and r. Thus, the point 's' lies within this circumcircle, if and only if its projection s onto the paraboloid lies within the lower half-space of the plane passing through p, q, r.

Now we can prove the main result.

Theorem: *Given a set of point S in the plane (assume no 4 are cocircular), and given three points p, q, r ϵ S, the triangle $\triangle pqr$ is a triangle to the Delaunay triangulation of S if and only if triangle $\triangle p'q'r'$ is a face of the lower convex hull of the projected set S'.*

From the definition of Delaunay triangulations we know that $\triangle pqr$ is in the

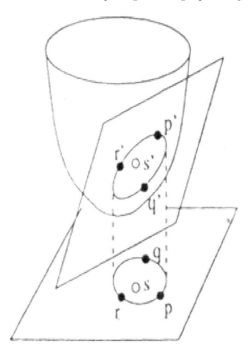

FIGURE 7.16
Planes and circles

Delaunay triangulation if and only if there is no point $s \epsilon S$ that lies within the circumcircle of pqr. From the previous lemma this is equivalent to saying that there is no point s' that lies in the lower convex hull of S', which is equivalent to saying that $p'q'r'$ is face of the lower convex hull. This completes the proof.

In order to test whether a point s lies within the circumcircle defined by p, q, r, it suffices to test whether s' lies within the lower half-space of the place passing through p', q', r'. If we assume that p, q, r are oriented counterclockwise in the plane this reduces to determing whether the quadruple p', q', r', s' is positively oriented, or equivalently whether s lies to the left of the oriented circle passing through p, q, r. This leads to the incircle test given by:

$$
In(p, q, r, s) = \det \begin{pmatrix} p_x & p_y & p_x{}^2 + p_y{}^2 & 1 \\ q_x & q_y & q_x{}^2 + q_y{}^2 & 1 \\ r_x & r_y & r_x{}^2 + r_y{}^2 & 1 \\ s_x & s_y & s_x{}^2 + s_y{}^2 & 1 \end{pmatrix} < 0 \qquad (7.16)
$$

7.11   Applications of Voronoi Diagram and Delaunay Triangulation

1. **Arrangements**

   The Voronoi diagram and Delaunay triangulations seem to be among the most important structures that can be derived from a set of points in the plane, when one considers the local structure of space. When one is interested in problems relating to the global structures of a set of points, it turns out that an altogether different structure is important, and this structure appears at first glance to have nothing to do with points, it has to do with line segments. An arrangement of a set of lines in the plane (or, generally, (d-1)-dimensional hyperplanes in d-dimensional space) is the graph whose vertices are the intersection points of the lines and whose edges are the line segments joining consecutive intersection points. The remarkable fact about line arrangements is that a large number of problems involving point sets can be solved by computing arrangements. The connection between the two is a concept of point-line duality, which allows us to translate problems on point sets into equivalent problems on sets of lines. Computing the line arrangement is a common first step in solving these problems.

2. **Search**

   Geometric search problems are of the following general form. Given a data set (e.g. points, lines, polygons) which will not change, preprocess this data set into a data structure so that some type of query can be answered as efficiently as possible. For example, a nearest neighbour search query is to determine the point of the data set that is closest to a given query point. A range query is to determine the set of points (or count the number of points) from the data set that lie within a given region. The region may be a rectangle, disc, or polygonal shape, like a triangle.

3. **Motion Planning and Visibility**

   Problems in this area include determining the shortest path between two points in the plane, given a set of polygonal obstacles that are to be avoided. This can be viewed as a geometric variant of the shortest path problem is graphs. It turns out that such shortest paths are made up of line segments that travel along lines of sight between obstacles, it is closely related to the problem of determining global visibility: namely, what objects can see what others.

7.11.1   Applications of Voronoi Diagrams

Voronoi diagrams have a number of important applications. These include:

1. **Nearest Neighbour Queries**
   One of the most important problems in computational geometry is solving nearest neighbour queries. Given a point set P, and given a query point q, determine the closest points in P to q. This can be answered by first computing a Voronoi diagram and then locating the cell of the diagram that contains q. This has many applications of pattern recognition and learning theory.

2. **Computational Morphology**
   Some of the most important operations in morphology are those of 'growing' and 'shrinking' (or 'thinning') objects. If we grow a collection of points, by imagining a grass fire starting simultaneously from each point, then the places where the grass fires meet will be along the Voronoi diagram. The medial axis of a shape (used in computer vision) is just a Voronoi diagram of its boundary. Region growing finds wider applications in computer vision.

3. **Finding Nearest Facility Location**
   If one want to establish a public utility facility such that, it should be placed as far as possible from any existing similar facilities. Where should it be placed? It turns out that the vertices of the Voronoi diagram are the points that locate at maximum distances from any other point in the set.

4. **High Clearance Path Planning**
   A robot wants to move around a set of obstacles avoiding collision. To minimize the possibility of collisions, it should stay as far away from the obstacles as possible. To do this, it should walk along the edges of the Voronoi diagram constructed with the 2D locations of the objects as the input domain.

7.12   Summary

A set of powerful tools for computing and analyzing the unstructured geometric data given in the form of a random set of points, lines, polygons and polyhedrons is defined as computational geometry. This chapter starts with the definition of some of the computational geometric primitives such as convex hull, Delaunay triangulation, Voronoi tessellation etc. The generic computational techniques to analyze the unstructured geometric data are discussed in terms of plane sweep method, randomized incremental algorithm and fractional cascading. The algorithms used to compute line-line intersection, convex hull computation, point inside a triangle, triangulation of simple polygon are discussed with their analysis. Further, the algorithms for computing the Delaunay triangulation and Voronoi tessellation are given with their properties and applications. Finally, applications of computational geometry in performing various functions of GIS are given with examples.

8

Spatial Interpolation Techniques

Spatial interpolation methods are important techniques for computation of spatial data at unsampled locations from sample data available in the domain. A number of methods have been developed for spatial interpolation in various disciplines and there are a number of different terms used to distinguish them, such as 'interpolating' and 'non-interpolating' methods or 'interpolators' and 'non-interpolators'. In this chapter, all these methods are referred to as spatial interpolation methods or spatial interpolators. Many of these methods have modified versions suitable to compute different variants of spatial data. The spatial interpolation methods covered in this chapter are only those commonly used in GIS and other spatial domains. As such, the list of the methods discussed in this chapter is not exhaustive. Only the frequently used methods and their variants are discussed.

In this chapter, important spatial interpolation methods used in GIS are described. They fall into three categories: (1) non-geostatistical methods, (2) geostatistical methods, and (3) combined methods. In geostatistics, the methods that are capable of using secondary information are often referred to as 'multivariate', while the methods that do not use the secondary information are called 'univariate' methods. Here it must be noted that multivariate usually refers to more than one response variable, despite of the fact that in some references it also refers to more than one explanatory variable (usually referred to as multiple variables). A brief introduction to geostatistics is provided prior to the descriptions of the geostatistical methods. The level of description of each method depends on the nature of the method.

Estimations of nearly all spatial interpolation methods can be represented as weighted averages of sampled data. They all share the same general estimation formula given by the equation

$$\hat{z}(x_0) = \sum_{i=1}^{n} \lambda_i z(x_i) \qquad (8.1)$$

where

$\hat{z}(x_0)$ is the estimated value of an attribute at the point of interest x_0

$Z(x_i)$ is the observed value at the sampled point x_i

λ_i, is the weight assigned to the sampled point

n represents the number of sampled points used for the estimation

8.1   Non-Geostatistical Interpolators

In this section twelve non-geostatistical interpolation methods are briefly described.

8.1.1   Nearest Neighbours

The nearest neighbours (NN) method predicts the value of an attribute at an unsampled point based on the value of the nearest sample by drawing perpendicular bisectors between sampled points (n), forming a structure such as Thiessen (Dirichlet or Voronoi) polygons ($V_i, i = 1, 2, .., n$). This produces one polygon per sample and the sample is located in the center of the polygon, such that in each polygon all points are nearer to its enclosed sample point than to any other sample points. The estimations of the attribute at unsampled points within polygon V_i are the measured value at the nearest single sampled data point x_i, that is $\hat{z}(xo) = z(x_i)$. The weights are given by the rule

$$if \ x_i \epsilon V_i \qquad\qquad (8.2)$$
$$than \ \lambda_i = 1 \ else \ \lambda_i = 0 \qquad\qquad (8.3)$$

All points (or locations) within each polygon are assigned the same value. A number of algorithms exist to generate the Thiessen polygons from a given set of points in plane.

8.1.2   Triangular Irregular Network

The triangular irregular network (TIN) was developed by Peuker and co-workers in 1978 for digital elevation modelling from an altitude matrix in the form of a grid. In TIN, all sampled points are joined into a series of triangles based on Delaunay's criteria known as Delaunay triangulation. The circumcircle of each of these triangles does not contain any other point from the sampled domain. This is known as the empty circumcircle property of Delaunay's triangulation. The TIN forms a different basis for making estimates in comparison with those used in nearest neighbour. The value of a point within a triangle is estimated by linear or cubic polynomial interpolation. The advantages and disadvantages of interpolation method using TIN are discussed in Burrough and McDonnell. The algorithm to generate Delaunay triangulation from a set of points in 2D plane is discussed in Chapter 7.

8.1.3   Natural Neighbours

The natural neighbours (NaN) method was introduced by Sibson (1981). It combines the best features of the nearest neighbour and TIN method. The

Non-Geostatistical	Geostatistical		Combined Method
	Univariate	Multivariate	
Nearest neighbours			Classification combined other interpolation methods
Triangular irregular network related	Simple kriging	Universal kriging	Trend surface analysis combined with kriging
Interpolations	Ordinary kriging	SK with varying local means	Lapse rate combined with kriging
Inverse distance weighting	Block kriging	Kriging with an external drift	Regression trees combined with kriging
Regression models	Factorial kriging	Simple cokriging	Residual maximum likelihood-empirical best linear
	Dual kriging	Ordinary cokriging	Regression kriging
Trend surface analysis		Ordinary cokriging	Regression kriging
Trend surface analysis		Standardised OCK	Gradient plus inverse distance squared
Splines and local trend surfaces	Indicator kriging	Principal component kriging	
Thin plate splines	Disjunctive kriging	Colocated cokriging	
Thin plate splines	Disjunctive kriging	Colocated cokriging	
Classification	Model-based kriging	Kriging within strata	
Regression tree	Simulation	Multivariate factorial kriging	
Fourier series		Indicator kriging	
Lapse rate		Indicator cokriging	
		Probability kriging	
		Simulation	

TABLE 8.1

The Spatial Interpolation Methods Considered in This Chapter first step in this method involves generation of a triangulation of the observed points using Delaunay's method, in which the apices of the triangles are the sample points in adjacent Thiessen polygons. This triangulation is unique except where the data are on a regular rectangular grid. To estimate the value of a point, it is inserted into the tessellation and then its value is determined by sample points within its bounding polygon. For each neighbour, the area of the portion of its original polygon that became incorporated in the tile of the new point is calculated. These areas are scaled to sum to 1 and are used as weights for the corresponding samples.

8.1.4 Inverse Distance Weighting

The inverse distance weighting or inverse distance weighted (IDW) method estimates the values of an attribute at unsampled points using a linear combination of values at sampled points weighted by an inverse function of the distance from the point of interest to the sampled points. The assumption is that sampled points closer to the unsampled point are more similar to it than those further away in their values. The weights can be expressed as:

$$\lambda_i = \frac{\frac{1}{d_i{}^p}}{\sum_{i=1}^{n} \frac{1}{d_i{}^p}} \tag{8.4}$$

where

d_i is the distance between x_0 and x_i.

p is a power parameter.

n represents the number of sampled points used for the estimation.

The main factor influencing the accuracy of IDW is the value of the power parameter. Weights diminish as the distance increases, especially when the value of the power parameter increases, so nearby samples have a higher weight and have more influence on the estimation, and the resultant spatial interpolation is local in nature than global.

The choice of power parameter and neighbourhood size is arbitrary. The most popular choice of p is 2 and the resulting method is often called inverse square distance or inverse distance squared (IDS). The power parameter can also be chosen on the basis of error measurement e.g. minimum mean absolute error, resulting in an optimal IDW. The smoothness of the estimated surface increases as the power parameter increases.

- IDW is referred to as "moving average" interpolation method when $p = 0$.
- IDW is referred as "linear interpolation" method when $p = 1$.
- IDW is referred as "weighted moving average" interpolation method when p is not equal to 1.

8.1.5 Regression Models

This method is essentially a linear regression model (LRM) and assumes that the data are independent of each other, normally distributed and homogeneous in variance. Regression methods explore a possible functional relationship between the primary variable and explanatory variables (e.g., geographical coordinates, elevation) that are easy to measure. These explanatory variables are usually referred to as secondary variables, auxiliary variables or ancillary variables. The information provided by these variables is called secondary information. The final model can be selected by a thorough understanding of the relationships between the primary variable and secondary variables or by using Bayesian information criteria (BIC).

8.1.6 Trend Surface Analysis

The trend surface analysis (TSA) is a special case of LRM, which only uses geographical coordinates to predict the values of the primary variable. TSA separates the data into regional trends and local variations. TSA shares the same assumption as LRM, and always contains all variables. It has also been extended to include other variables, in which case, it should be classified as LRM.

8.1.7 Splines and Local Trend Surfaces

The splines consist of polynomials with each polynomial of degree p being local rather than global. The polynomials describe pieces of a line or surface (i.e. they are fitted to a small number of data points exactly) and are fitted together so that they join smoothly. The places where the pieces join are called knots. The choice of knots is arbitrary and may have a dramatic impact on the estimation. For degree p= 1, 2, or 3, a spline is called linear, quadratic or cubic respectively. Typically the splines are of degree 3 and they are cubic splines.

The local trend surfaces (LTS) fit a polynomial surface for each predicted point using the nearby samples. There are two approaches in LTS. The first is a local polynomial regression fitting that is detailed by Cleveland et al. and Cleveland and Devlin (1988). The second is a bilinear or bicubic spline that was developed to implement bivariate interpolation onto a grid for irregularly spaced point data. This method is also known as Akima's interpolator (AK). Both approaches are unable to choose the smoothness.

8.1.8 Thin Plate Splines

Thin plate splines (TPS), formally known as "laplacian smoothing splines", were developed principally by Wahba and Wendelberger (1980) for interpolation of climatic data. The smoothing parameter is calculated by minimising the generalised cross validation function (GCV). This method is relatively robust because the minimisation of GCV directly addresses the predictive accuracy and is less dependent on the veracity of the underlying statistical model. TPS provides a measure of spatial accuracy.

8.1.9   Classification Methods

The classification method uses easily accessible soft information (e.g., soil types, vegetation types, or administrative areas) to divide the region of interest into sub-regions that can be characterised by the moments (i.e., mean, variance) of the attribute measured at locations within the region of interest (Burrough and McDonnell [8]). The model for classification is:

$$\hat{z}(x_0) = \mu + \alpha_k + \epsilon \tag{8.5}$$

where

$\hat{z}$ is the estimated value of the attribute at location x_0.

μ is the general mean of the attribute over the region of interest.

α_k is the deviation between μ and the mean of type k.

ϵ is the residual error.

The class of the sample can be computed using the analysis of variance method or LM by specifying the attribute as a response variable and the soft information as an explanatory factor with k classes. This method shares the same assumptions as LM.

8.1.10   Regression Tree

The regression tree, also known as binary decision trees, uses binary recursive partitioning whereby the data of the primary variable are successively split along the gradient of the explanatory variables into two descendent subsets or nodes. These splits occur so that at any node the split is selected to maximise the difference between two split groups or branches. The mean value of the primary variable in each terminal node can then be used to map the variable across the region of interest.

8.1.11   Fourier series

The Fourier series (FS) method is used to estimate the values of an attribute by interpolating the samples using a linear combination of sine and cosine waves in two-dimensional space, as follows:

$$
\hat{Z} = \sum_{m=1}^{\infty} \sum_{n=1}^{\infty} \alpha_{nm} \cos \frac{2n\pi X_i}{\lambda_x} \cos \frac{2n\pi Y_i}{\lambda_y} + \sum_{m=1}^{\infty} \sum_{n=1}^{\infty} \beta_{nm} \cos \frac{2n\pi X_i}{\lambda_x} \sin \frac{2n\pi Y_i}{\lambda_y}
$$
$$
+ \sum_{m=1}^{\infty} \sum_{n=1}^{\infty} \gamma_{nm} \sin \frac{2n\pi X_i}{\lambda_x} \cos \frac{2n\pi Y_i}{\lambda_y} + \sum_{m=1}^{\infty} \sum_{n=1}^{\infty} \delta_{nm} \sin \frac{2n\pi X_i}{\lambda_x} \sin \frac{2n\pi Y_i}{\lambda_y}
$$

$$(8.6)$$

8.1.12 Lapse Rate

The lapse rate (LR) was developed to estimate air temperature with respect to elevation/altitude. It uses the temperature value of the nearest weather station and the difference in elevation to estimate air temperature at the unsampled point on the basis of the relationship between air temperature and elevation for a region. It is also termed smart interpolation. It makes the assumption that the lapse rate is constant across the study region. Several variants of LR have been proposed for air temperature, given it is limited to only predicting temperature using elevation.

8.2 Geostatistics

In this section twelve geostatistical interpolation methods are briefly described.

8.2.1 Introduction of Geostatistics

Geostatistics is usually believed to have originated from the work in geology and mining by Krige (1951), but it can be traced back to the early 1910s in agronomy and 1930s in meteorology. It was developed further by Matheron (1963) with his theory of regionalised variables." A mineralized phenomenon can be characterized by the spatial distribution of a certain number of measurable quantities called regionalized variables," and this concept is termed regionalisation. Other key concepts of geostatistics are: When a variable is distributed in space, it is said to be regionalized and geostatistical theory is based on the observation that the variabilities of all regionalized variables have a particular structure. Geostatistics includes several methods that use kriging algorithms for estimating continuous attributes. Kriging is a generic name for a family of generalised least-squares regression algorithms, named in recognition of the pioneering work of Danie Krige (1951).

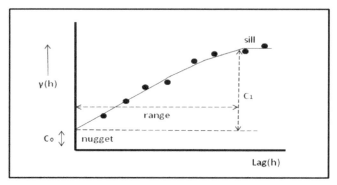

Variogram with range, nugget, and sill

FIGURE 8.1
Variogram with range, nugget and sill

8.2.2   Semivariance and Variogram

Semivariance γ of Z between two data points is an important concept in geostatistics and is defined as:

$$\gamma(x_i, x_0) = \gamma(h) = \frac{1}{2}var[Z(x_i) - Z(x_0)] \qquad (8.7)$$

where h is the distance between point x_i and x_0 and $\gamma(h)$ is the semivariogram commonly referred to as variogram.

A plot of x_i and x_0 and $\gamma(h)$ against h is known as the experimental variogram, which displays several important features. Various parameters that can be derived from a semi variogram is depicted in Figure 8.1. The "nugget", is a positive value of $\gamma(h)$ as h close to 0, which is the residual reflecting the variance of sampling errors and the spatial variance at shorter distance than the minimum sample spacing. The "range" is a value of distance at which the "sill" is reached. Samples separated by a distance larger than the range are spatially independent because the estimated semivariance of differences will be invariant with sample separation distance. If the ratio of sill to nugget is close to 1, then most of the variability is non-spatial. The range provides information about the size of a search window used in the spatial interpolation methods.

The semivariance can be estimated from the data using the equation

$$\hat{\gamma}(h) = \frac{1}{2n} \sum_{i=1}^{n} (z(x_i) - z(x_i + h))^2 \qquad (8.8)$$

where n is the number of pairs of sample points separated by distance h.

Variogram modelling and estimation is extremely important for structural analysis and spatial interpolation. The variogram models may consist of simple models, including nugget, exponential, spherical, Gaussian, linear, and power model or the nested sum of one or more simple models. Four commonly used variogram models are illustrated in Figure 8.2.

8.2.3 Kriging Estimator

All kriging estimators are variants of the basic equation, which is a slight modification of equation, as follows:

$$\hat{Z}(x_0) - \mu = \sum_{i=1}^{n} \lambda_i [Z(x_i) - \mu(x_0)] \tag{8.9}$$

where

μ is a known stationary mean, assumed to be constant over the whole domain and calculated as the average of the data

λ_i is kriging weight

n is the number of sampled points used to make the estimation and depends on the size of the search window

$\mu(x_0)$ is the mean of samples within the search window

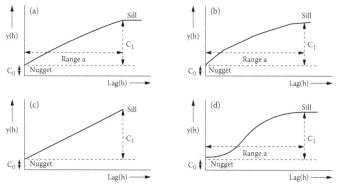

Commonly used Variogram models: (a) Spherical; (b) Exponential; (c) Linear; and (d) Gaussian

FIGURE 8.2
Commonly used variogram models: (a) spherical; (b) exponential; (c) linear; and (d) Gaussian

The kriging weights are estimated by minimising the variance, as follows:

$$Var[\hat{Z}(x_0)] = E[\hat{Z}(x_0) - z(x_0)^2] \quad (8.10)$$

$$Var[\hat{Z}(x_0)] = E[\hat{z}(x_0)^2 + Z(x_0)^2 - 2\hat{z}(x_0)z(x_0)] \quad (8.11)$$

$$Var[\hat{Z}(x_0)] = \sum_{i=1}^{n}\sum_{j=1}^{n}\lambda_i\lambda_j C(x_i - x_j) + C(x_0 - x_0) - 2\sum_{i=1}^{n}\lambda_i C(x_i - x_0) \quad (8.12)$$

where

$Z(x_0)$ is the true value expected at point x_0

n represents the number of observations to be included in the estimation

$C(x_i - x_j) = Cov[Z(x_i), Z(x_j)]$ is the covariance matrix

The assumptions of kriging are stationary of difference between x and x+h and variance of difference, which define the requirements for the intrinsic hypothesis. This means that semivariance does not depend on the location of samples but depends only on the distance between samples, thus semivariance is isotropic in nature.

Kriging is a generic method for many geostatistical interpolation techniques. Geostatistical approaches are used to describe spatial patterns and interpolate the value of the primary variable at unsampled locations. Also it models the uncertainty of the error of the estimated surface. There are many variants of kriging, depending upon the models. The variants are discussed below.

8.2.4   Simple Kriging

The estimation of simple kriging (SK) is a modified variant of the kriging estimator is given by the equation

$$\hat{Z}(x_0) = \sum_{i=1}^{n}\lambda_i Z(x_i) + [1 - \sum_{i=1}^{n}\lambda_i]\mu \quad (8.13)$$

where μ is a known stationary mean. The parameter μ is assumed constant over the entire domain and calculated as the average of the data. It is also known as kriging with known mean because it is used to estimate residuals from the reference value μ given apriori. Here the parameter $\mu(x_0)$ of the kriging estimator is replaced by the stationary mean μ. Simple kriging does not have a non-bias condition, $1 - \sum_{i=1}^{n}\lambda_i$ is not necessarily equal to zero. The greater the value of $1 - \sum_{i=1}^{n}\lambda_i$, the estimator will move toward the mean and in general the value of $1 - \sum_{i=1}^{n}\lambda_i$ is relatively poor in the sampled regions. SK assumes second-order stationary that is constant mean, variance and covariance over the domain or the region of interest. Because of such an assumption it is often too restrictive.

8.2.5 Ordinary Kriging

Ordinary kriging (OK) is similar to SK and the only difference is that OK estimates the value of the attribute using equations 8.9 and 8.13 by replacing μ with a local mean $\mu(x_0)$ that is the mean of samples within the search window, and forcing $[1 - \sum_{i=1}^{n} \lambda_i] = 0$ that is $\sum_{i=1}^{n} \lambda_i = 1$, which is achieved by plugging it into equation 8.13. OK estimates the local constant mean, then performs SK on the corresponding residuals, and only requires the stationary mean of the local search window.

8.2.6 Kriging with a Trend

Kriging with a trend (KT) is normally called universal kriging (UK) and was proposed by Matheron in 1969. It is an extension of OK by incorporating the local trend within the neighbourhood search window as a smoothly varying function of the coordinates. UK estimates the trend components within each search neighbourhood window and then performs simple kriging on the corresponding residuals.

8.2.7 Block Kriging

Block kriging (BK) is a generic name for estimation of average values of the primary variable over a segment, a surface, or a volume of any size or shape. It is an extension of OK and estimates a block value instead of a point value by replacing the point-to-point covariance with the point-to-block covariance. Essentially, BK is block ordinary kriging and ordinary kriging is "point" ordinary kriging.

8.2.8 Factorial Kriging

Factorial kriging (FK) is designed to determine the origins of the value of a continuous attribute. It models the experimental semivariogram as a linear combination of a few basic structure models to represent the different factors operating at different scales (e.g., local and regional scales). FK can decompose the kriging estimates into different components such as nugget, short-range, long-range and trend, and such components could be filtered in mapping if considered as noise. For example, the nugget component at sampled points could be filtered to remove discontinuities (peaks) at the sampled points, while the long-range component could be filtered to enhance the short-range variability of the attribute. FK assumes that noise and the underlying signal are additive.

8.2.9 Dual Kriging

Dual kriging (DuK) estimates the covariance values instead of data values to elucidate the filtering properties of kriging. It also reduces the computational cost of kriging when used with a global search neighbourhood. It includes dual SK, dual OK, and dual FK. Its uses are restricted to some specific applications.

8.2.10 Simple Kriging with Varying Local Means

Simple kriging with varying local means (SKVLM) is an extension of SK by replacing the stationary mean with known varying means at each point that depends on the secondary information. If the secondary variable is categorical, the primary local mean is the mean of the primary variable within a specific category of the secondary variable. If it is continuous, the primary local mean is a function of the secondary variable or can be acquired by discretising it into classes. SK is then used to produce the weights and estimates.

8.2.11 Kriging with an External Drift

Kriging with an external drift (KED) is similar to UK but incorporates the local trend within the neighbourhood search window as a linear function of a smoothly varying secondary variable instead of as a function of the spatial coordinates. The trend of the primary variable must be linearly related to that of the secondary variable. This secondary variable should vary smoothly in space and is measured at all primary data points and at all points being estimated. KED is also called UK or external drift kriging in Pebesma (2004). KED could be extended to include both secondary variables and coordinate information if gstat is used.

8.2.12 Cokriging

Unlike SK within strata (see Section 2.3.15), SKlm and KED that require the availability of information of secondary variables at all points being estimated, cokriging (CK) is proposed to use non-exhaustive secondary information and to explicitly account for the spatial cross correlation between the primary and secondary variables. Equation 8.13 can be extended to incorporate the additional information to derive equation for CK as follows:

$$\hat{Z}_1(x_0) - \mu_1 = \sum_{i_1=1}^{n_1} \lambda_{i_1} [Z_1(x_{1_i}) - \mu_1(x_{i_1})] + \sum_{j=2}^{n_i} \sum_{i_j} = 1^{n_j} \lambda_{i_j} [Z_j(x_{i_j}) - \mu_j(x_{i_j})]$$

(8.14)

where

μ_i is a known stationary mean of the primary variable

$z_j(x_{ij})$ is the data of the primary variable

$\mu_j(x_{ij})$ is the mean of samples within the search window

n_i is the number of sampled points within the search window for point x_0 used to make the estimation

λ_{i_1} is the weight selected to minimize the estimation variance of the primary variable

n_i is the number of secondary variables

n_j is the number of jth secondary variable within the search widow

ref is the weight assigned to (ref) point of h secondary variable, Z (xi) is the data at ref point of ref secondary variable, and is the mean of samples of ref secondary variable within the search window

The cross-semi variance (or cross-variogram) can be estimated from data using the following equation 8.15:

$$\hat{\gamma}_{12}(h) = \frac{1}{2n} \sum_{i=1}^{n} [z_1(x_i) - z_1(x_i + h)][z_2(x_i) - z_2(x_i + h)] \qquad (8.15)$$

where n is the number of pairs of sample points of variable z_j and z_2 at point x, $x_i + h$ separated by distance h. Cross-semivariances can increase or decrease with h depending on the correlation between the two variables and the Cauchy-Schwarz relation must be checked to ensure a positive CK estimation variance in all circumstances.

8.3 Summary

This chapter discusses the important methods of spatial interpolation. It starts with the definition of spatial interpolation as a technique for computation of spatial data at unsampled locations from given sampled data. It classifies the spatial interpolation techniques available so far into two categories non-geostatistical interpolators and geostatistical interpolators. Among the prominent non-geostatistical interpolators discussed are the nearest neighbour, triangular irregular network (TIN), natural neighbour, inverse distance weighting (IDW), regression methods, trend surface analysis, thin plate splines, classification techniques and Fourier series methods. The geostatistical methods start with introduction of variance and semi variogram. The methods discussed under this category are variogram method, Kriging, and its variants such as ordinary kriging, block kriging, factorial kriging, kriging with external drift and cokriging. The application of these interpolation techniques to spatial data from different domain are discussed at the end of the chapter. This chapter gives good exposure to different spatial interpolation methods through a clear definition, mathematical explanation and applications in various application domains.

9

Spatial Statistical Methods

Statistics plays an important role in describing the sample data and exploring the behavior of population behavior from sample data. The traditional statistical methods have been developed keeping in view describing the behavior of univariate data sample. To incorporate the behavior of spatial data and geometric data which are inherently univariate or multivariate in nature, a set of tools has been developed under the aegis of spatial statistics or geostatistics. In this chapter we discuss the spatial statistical methods and their usage. The standard statistical tools useful for describing the behavior of univariate data sample are discussed.

GIS is a computing system capable of collecting, collating and processing samples of spatial data to visualize and analyze the patterns present in the spatial data population through a set of descriptive and inferential statistical methods.

9.1 Definition of Statistics

Statistics is the study of how to collect, organizes, analyze, and interpret numerical information from data. Descriptive statistics involves methods of organizing, picturing and summarizing information from data. Inferential statistics involves methods of using information from a sample to draw conclusions about the data population.

Statistical inferences are as accurate as the data samples they are applied to. Statistical results should be interpreted by one who understands the methods used as well as the subject matter. Statistical methods are so widely used today that it is difficult to enumerate the various spheres of their application; statistics is used in every department of government and industry.

Statistics deals with aggregates of objects and does not take cognizance of individual items. For example, in finding the statures of a class, a statistician is not very much interested in the height of individual students but in their average height. It is immaterial for him whether a particular student is five feet or seven feet. One wants to have a bird's-eye view, so to say, of their height perhaps by way of comparing it with the average heights of students in the same class at some other college or in a different class at the same institution. Statistical laws are not exact. The results of statistical enquiry are not expressed in the form of categorical certainty but in terms of probabilities only. A statistical enquiry passes through various stages as given below:

- Collection of sample data or field survey.
- Organization and preprocessing of survey data and classification or categorization of the sample data.
- Analysis of the data using statistical operators and methods.
- Interpretation of the patterns and trends in the data.

9.2 Spatial Statistics

Spatial statistics can be defined as a set of analytical techniques to determine the spatial distribution of a variable, the relationship between the spatial distribution of variables, and the association of the variables of a spatial extent or area. Spatial analysis is often referred to as spatial modeling. It refers to the analysis of phenomena distributed in space and having physical dimensions (the location of, proximity to, or orientation of objects with respect to one another; relating to an area of a map as in spatial information and spatial analysis; referenced or relating to a specific location on the Earth's surface).

Spatial analysis is the process of extracting or creating new information from a set of spatial features to perform assessment, evaluation, analysis or modeling of data in a spatial extent or area based on pre-established and computerized criteria and standards. Spatial analysis is a process of modeling, examining, and interpreting model results useful for evaluating suitability and capability, for estimating and predicting, and for interpreting and understanding of spatial phenomena.

Generally there are four traditional types of spatial analysis practiced in GIS. They are spatial overlay analysis and contiguity analysis, surface analysis, linear analysis, and raster analysis. Spatial analysis includes GIS functions such as topological analysis of overlays using computational geometric queries, buffer generation around spatial objects and filtering of a specific type of spatial object within the buffer zone, and spatial or network modeling and analysis of shortest, optimum and alternate routes between chosen source and destination. Spatial analysis also includes analysis of spatial patterns and spatial relationships among objects and events.

Spatial statistics can be seen as a set of computing methods performing the following functions:

- Analysis of point pattern in 2D and 3D space.
- Computing spatial autocorrelation of objects.
- Computing correlation and regression of spatial phenomena.

9.3 Classification of Statistical Methods

Based on the type of data they operate and inference they draw, statistical methods can be classified into (a) descriptive statistical method and (b) inferential statistical method. Generally descriptive statistics operate of univariate sample set of data and describe the behavior through a quantity derived from the set. Inferential statistical methods operate on multivariate samples and show the correlation and relationship among various parameters of spatial phenomena. Before explaining the significance of the univariate, descriptive statistical methods, it is pertinent to have a fair understanding of the difference between univariate and bivariate data, and descriptive and inferential statistics. Table 9.1 delineates the difference between univariate and bivariate data.

Univariate analysis explores each variable in a data set separately. It looks at the range of values, as well as the central tendency of the values. It describes the pattern of response to the variable. It describes each variable on its own. Descriptive statistics describe and summarize data. Univariate descriptive statistics describe individual variables.

9.3.1 Descriptive Statistics

Descriptive statistics is the discipline of quantitatively describing the main features of a collection of data. Descriptive statistics are distinguished from inferential statistics (or inductive statistics) in that descriptive statistics aim to summarize a sample, rather than use the data to learn about the population that the sample of data is thought to represent.

Descriptive statistics provides simple summaries about the sample and about the observations that have been made. Such summaries may be either quantitative, i.e. summary statistics, or visual, i.e. simple-to-understand graphs. These summaries may either form the basis of the initial description of the data as part of a more extensive statistical analysis, or they may be sufficient in and of themselves for a particular investigation.

In statistics, statistical inference is the process of drawing conclusions from data that is subject to random variation, for example, observational errors or sampling variation. More substantially, the terms statistical inference, statistical induction and inferential statistics are used to describe systems of procedures that can be used to draw conclusions from data sets arising from systems affected by random variation, such as observational errors or random sampling.

Univariate Data	Bivariate Data
• Data involving a single variable.	• Data involving two variables.
• Does not deal with causes or relationships.	• Deals with causes or relationships.
• The major purpose of univariate analysis is to describe.	• The major purpose of bivariate analysis is to explain.
• Central tendency of data is measured using mean, mode, median. • The dispersion or spread of the data is measured through range, variance, max, min, quartiles, standard deviation. • The frequency distributions is described through bar graph, histogram, pie chart, line graph, box-and-whisker plot.	• Analysis of two variables simultaneously. • Correlations are used to understand the relationship between the variables. • Bivariate analysis leads to comparison between attributes, discovers emperical relationships among correlated attributes, explains cause and effect relationship among variables. • Discovers the independent and dependent variables.
Sample Question: How many of the students in the freshman class are female?	**Sample Question:** Is there a relationship between the number of females in computer programming and their scores in mathematics?

TABLE 9.1
Comparison of Univariate and Bivariate Data

9.4   Role of Statistics in GIS

The following example will show how an emerging set of tools that rely on spatial statistics provides GIS users the capability to conduct spatial analysis of the information we have.

1. **Calculating the center, dispersion, and trend**

   Statistics can describe the characteristics of a set of features, including the center of the features and the extent to which features are clustered or dispersed around the center, and any directional trend. Analyzing the distribution of features is useful for studying change over time for example, to see if the center of cases of a particular disease changes position over the course of several months, or for computing two or more sets of features.

2. **Identifying patterns in spatial data**

   We can use spatial statistics to measures whether and to what extent the distribution of features creates a pattern.

   For example, extracting the prevalence and spread of a particular disease like malaria or water born disease in an area. Depending on the spread the hygienic condition or the source of the water-borne disease can be identified in the area. If we find the classes of disease are clustered, the source is likely somewhere inside the cluster, such as a pond harbouring infected mosquitoes.

   We can also identify patterns in the distribution of attribute values associated with the features. For example, we might calculate the degree to which student test scores in a city are clustered. If similarly high or low scores occur together, it may mean money and other resources are affecting the scores.

3. **Identifying spatial clusters**

   Here we discuss the importance of identifying clusters. We can determine if the features or values associated with the features occur together, and measure the strength of the relationship. For example, a public health analyst could determine the extent to which economic or demographic factors are related to the quality of infant health in neighbourhoods across a country. We can use relationships to make predictions about where features or certain attribute values will osculate action or when we want to find the cause of the cluster, so we know what action to take. A public health department would take immediate action to notify people living a flu cluster has been identified to watch for symptoms. They could then try to identify the source of the outbreak—if it's a school, they would know to begin inoculating the children. We can also use statistics to identify clusters of features with similar values. For example, a tax assessor could create neighbourhoods by identifying clusters of block groups with similar median house values.

4. **Geo-statistics** is a branch of statistics focusing on spatial or spatio-temporal data sets.

5. **Spatial analysis or spatial statistics**
This includes any of the formal techniques which study entities using their topological, geometric, or geographic properties. It refers to a variety of techniques, many still in their early development, using different analytic approaches and applied in diverse fields.

9.5   Descriptive Statistical Methods

Some of the frequently computed descriptive statistical methods are:

Mean

Median

Mode

Variance

Standard deviation

Standard error

Range

Skewness

Kurtosis

For univariate data population $Y_1, Y_2, .., Y_N$, the statistical quantity such as mean, mode, median, variance, standard deviation, best estimation of standard deviation, mean deviation, range, skewness and kurtosis are some of the statistical parameters which give quantitative trend measures of the sample population.

9.5.1   Mean

The sum of a list of numbers, divided by the total number of numbers in the list is the mean. The arithmetic mean of a set of values is the quantity commonly called the mean or the average. Given a set of samples, the arithmetic mean is given by equation 9.1.

$$\hat{Y} = \mu = \frac{1}{N} \sum_{i=1}^{N} Y_i \tag{9.1}$$

where $\hat{Y} = \mu$ is the mean, N is the sample size, and Y_i is the ith sample.

9.5.2 Median

Median is the *middle value of the ordered list of values*—the smallest number such that at least half the numbers in the list are no greater than it. If the list has an odd number of entries, the median is the middle entry in the list after sorting the list into increasing order. If the list has an even number of entries, the median is equal to the sum of the two middle numbers divided by two.

9.5.3 Mode

The mode is *the value that appears most often in a set of data*. Like the statistical mean and median, the mode is a way of expressing, in a single number, important information about a random variable or a population. The numerical value of the mode is the same as that of the mean and median in a normal distribution, and it may be very different in highly skewed distributions. The mode is not necessarily unique, since the same maximum frequency may be attained at different values. The most extreme case occurs in uniform distributions, where all values occur equally frequently.

9.5.4 Variance

In statistics, in a population of samples, *the mean of the squares of the differences between the respective samples and the sample mean* is called variance. It is expressed mathematically as equation 9.2.

$$\sigma^2 = \frac{1}{N} \sum_{i=1}^{N} (Y_i - \hat{Y})^2 \tag{9.2}$$

where σ^2 is the standard deviation, $\hat{Y}$ is the sample mean, N is the sample size and Y_i is the ith sample.

9.5.5 Standard Deviation

The *square root of the sample variance of a set of N values* is called standard deviation (SD) of the sample and given by equation 9.3.

$$S_N = \sqrt{\frac{1}{N} \sum_{i=1}^{N} (Y_i - \hat{Y})^2} \tag{9.3}$$

9.5.5.1   Best Estimation of Standard Deviation

The square of the sample variance of a set of N values is called standard deviation. By best estimation, we mean that the estimator should be unbiased. Hence the formula for best estimation is given by the equation 9.4.

$$S^2 = \frac{1}{N-1}[\sum_{i=1}^{N} Y_i^2 - N * \hat{Y}_N^2] \qquad (9.4)$$

where S is the mean square for the sample, N is the number of the samples, $\hat{Y}$ is the mean.

9.5.5.2   Mean Deviation

The mean deviation is the *mean of the absolute deviation of a set of data about the N data's mean.* For the sample size N having $\hat{Y}$ as the mean of the distribution the mean deviation is given by equation 9.5.

$$MD = \frac{1}{N}\sum_{i=1}^{N}|Y_i - \hat{Y}| \qquad (9.5)$$

9.5.6   Standard Error

The standard error of a statistic is the standard deviation of the sampling distribution of the statistic. The standard error of a statistic depends on the sample size. In general, the larger the sample's size the smaller is the standard error. Standard errors are important because they reflect how much sampling fluctuation a statistic will show. The formula for the standard error of the mean is given by the equation 9.6

$$\sigma_M = \frac{\sigma}{\sqrt{N}} \qquad (9.6)$$

where σ is the standard deviation of the original distribution and N is the sample size.

9.5.7   Range

Range is *the difference between the greatest and the least value of the variant.* It is easy to calculate it and it gives a general idea about the distribution. Its use is very much limited as it does not take into account the central tendency of the form of the distribution. Note that *the range is a single number, not many numbers.*

9.5.8 Skewness

Skewness is *the lack of symmetry* and is termed as positive if the longer tail of the frequency curve is towards the higher value of variant. The skewness for a normal distribution is zero and any symmetric data should have skewness near zero.

- Negative value for the skewness indicates that data are skewed left. By skewed left, we mean that the left tail is heavier than the right tail.
- Positive value for the skewness indicates that data are skewed right. By skewed right, we mean that the right tail is heavier than the left tail.

For univariate data $Y_1, Y_2, .., Y_N$ the formula for skewness is given by equation 9.7.

$$Skewness = \frac{\sum_{i=1}^{N}(Y_i - \hat{Y})^3}{(N-1)^3} \tag{9.7}$$

where $\hat{Y}$ is the mean, S is standard deviation and N is the number of data points in the sample population.

9.5.9 Kurtosis

Kurtosis is *a measure of whether the data are peaked or flat relative to a normal distribution*, i.e. data sets with high kurtosis tend to have a distinct peak near the mean, decline rather rapidly, and have heavy tails. Data sets with low kurtosis tend to have a flat top near the mean rather than a sharp peak. A uniform distribution would be the extreme case.

Kurtosis is based on the size of a distribution's tails. Distributions with relatively large tails are called 'leptokurtic', those with small tails are called 'platykurtic'. A distribution with the same kurtosis as the normal distribution is called 'mesokurtic'. The standard normal distribution has a kurtosis of zero. Positive kurtosis indicates a 'peaked' distribution and negative kurtosis indicates a 'flat' distribution.

For univariate data $Y_1, Y_2, .., Y_N$ the formula for kurtosis is given by equation 9.8.

$$kurtosis = \frac{\sum_{i=1}^{N}(Y_i - \hat{Y})^4}{(N-1)^4} \tag{9.8}$$

where $\hat{Y}$ is the mean, S is standard deviation and N is the number of data points in the sample population.

9.6   Inferential Statistics

Description and descriptive statistics deals with obtaining summary measures to describe a set of data, whereas inference and inferential statistics are concerned with making inferences about the population from the surveyed samples. Inferential statistics is concerned with making legitimate inferences about underlying processes from observed patterns. Therefore inferential statistics is often used for sample space where obtaining data through survey and observation is difficult. Applications such as astronomy, radiation monitoring, natural disaster use inferential statistics to study the observations and predict about the phenomena.

In most situations, there is a lack of availability of data for an entire population with all possible occurrences. Most statistical measures are estimated based on sample data and calculated from samples which are estimates of population parameters. Sampling is carried out by human beings measuring the parameter or measured through sensors. Therefore, there is always an element of chance or error in the sampling process. Sometimes two different statistical measures on the same sample may yield different results. Then the question arises whether an observed difference (say between two statistics) could have arisen due to chance associated with the sampling process, or reflects a real difference in the underlying population(s). The answers to this question involve the concepts of statistical inference and statistical hypothesis testing. It is always important to explore before any firm conclusions are drawn. However, before taking any decision based on statistical significances, it should be kept in mind that, statistical significance does not always equate to scientific (or substantive) significance. In GIS often the data size is large and the sample size is big enough. Therefore statistical significance is easily achieved for spatial data in GIS.

The inferential statistical tools discussed in this section are correlation coefficient, weights matrix, join count statistics. The statistics of correlated spatial characteristics are inferred using Moran's I, Geary's C, and General G.

9.6.1   Correlation Coefficient (R)

Correlation and regression of two variables are defined for both standard variables and spatial variables. Correlation coefficient is one of the classic descriptive Statistics measure for bivariate variables. This is also known as Pearson Product Moment Correlation Coefficient (PPMCC) and designated as R. The R measures the degree of association or strength of the relationship between two continuous variables. The computation of R for a bivariate sample population is carried out through equation 9.9.

$$R = \frac{\sum\limits_{i=1}^{n}(x_i - \bar{X})(y_i - \bar{Y})}{nS_xS_y} \tag{9.9}$$

where

$$S_x = \sqrt{\frac{\sum\limits_{i=1}^{n}(x_i-\bar{X})^2}{N}}$$

and $S_y = \sqrt{\frac{\sum\limits_{i=1}^{n}(y_i-\bar{Y})^2}{N}}$

The S_x and S_y are standard deviations in x and y directions respectively.

The computed value of R for any data population varies on a scale from $+1$ through 0 to -1.

- $R = +1$ implies perfect positive association, whereby as values of one rise they also rise of the other. The phenomena of increase in real estate rate in a locality implies high and increased income of occupants and increase per capita income of the locality.
- $R = 0$ implies no association between the phenomena or attribute under study. For example the annual rate of rainfall over an area has no correlation to the education index of the locality.
- $R = -1$ implies perfect negative association, whereby as values of one variable rise, those of the other fall. A perfect example of negative association can be that as the price of commodity increases the quantity purchased by the consumer decreases. If the education level of human population in a locality increases the rate of crime decreases.

9.6.2 Moran's Index, or Moran's I

Moran's I is a measure of spatial autocorrelation developed by Patrick A. P. Moran. Spatial autocorrelation is characterized by a correlation in a signal among nearby locations in space. It is more complex than one-dimensional autocorrelation because is multi-dimensional (i.e. 2 or 3 dimensions of space) and multi-directional.

Moran's I is defined by equation 9.10

$$I = \frac{N}{\sum\limits_{i}\sum\limits_{j}w_{ij}} \frac{\sum\limits_{i}\sum\limits_{j}w_{ij}(X_i - \hat{X})(X_j - \hat{X})}{\sum\limits_{i}(X_i - \hat{X})^2} \tag{9.10}$$

where N is the number of spatial units measured at locations indexed by i and

j. X is the variable of interest, $\hat{X}$ is the mean of X_i and w_{ij} is an element of a matrix of spatial weights.

Moran's I is applied to a continuous variable of polygons or points. It is similar to correlation coefficient R and its value varies between $-1.0 \, and +1.0$ through 0.

1. $I = 0$ or close to zero [technically, close to $\frac{-1}{n-1}$], indicate a random sampling pattern or no spatial autocorrelation among the observations.

2. $I = +1$ or indices above $\frac{-1}{n-1}$ (towards +1) indicate a strong tendency toward clustering of spatial observations; autocorrelation is high and positive.

3. $I = -1$ or indices below $\frac{-1}{n-1}$ (toward -1) indicate a tendency toward dispersion or spread uniformly with no clustering or autocorrelation is negative.

Therefore Moran's Index is used for inferring the dispersion, randomness or cluster patterns of a spatially varying point patterns. Moran's I, differs from correlation coefficient R in the following aspects:

1. It involves one variable only, not two variables.

2. It incorporates weights matrix (w_{ij}), which is an index of relative location of the observations.

3. It may be seen as the correlation between neighbouring values on a variable.

4. The correlation between variable X and the spatial lag of X is formed by averaging all the values of X for the neighbouring samples.

9.6.3   Geary's C

Geary's contiguity ratio is also known as Geary's C, Geary's ratio, or the Geary's index. This statistic was developed by Roy C. Geary. Geary's C is a measure of spatial autocorrelation. Like autocorrelation, spatial autocorrelation means that adjacent observations of the same phenomenon are correlated. However, autocorrelation is about proximity in time. Spatial autocorrelation is about proximity in (two-dimensional) space. Spatial autocorrelation is more complex than autocorrelation because the correlation is two-dimensional and bi-directional.

Geary's C is defined by equation 9.11:

$$C = \frac{(N-1)\sum_i \sum_j w_{ij}(X_i - X_j)^2}{2W \sum_i (X_i - \hat{X})^2} \qquad (9.11)$$

where N is the number of spatial units indexed by i and j; X is the variable of interest; $\hat{X}$ is the mean of X; w_{ij} is a matrix of spatial weights; W is the sum of all w_{ij}.

The value of Geary's C lies between $[0, 2]$.

1. Geary's $C = 1$ means no spatial autocorrelation and the variable is spatially random.

2. Geary's $C = 0$ indicates perfect positive autocorrelation and the data is spatially clustered.

3. Geary's $C = 2$ indicates perfect negative autocorrelation and data is spatially spread.

Values lower than 1 demonstrate increasing positive spatial autocorrelation, whereas values higher than 1 indicate increasing negative spatial autocorrelation. In general,

1. If $(0 < C < 1)$ then the clustered pattern in which adjacent points show similar characteristics.

2. If $(C \approx 1)$ then the random pattern in which points do not show particular patterns of similarity.

3. If $(1 < C < 2)$ then the pattern is dispersed / uniform pattern in which adjacent points show different characteristics.

Geary's C is inversely related to Moran's I, but they are not identical. Moran's I is a measure of global spatial autocorrelation and is applied to observations spread globally, while Geary's C is more sensitive to local spatial autocorrelation and is applied to closely surveyed samples.

9.6.4 General G Statistic

Moran's I and Geary's C are used to indicate clustering or positive spatial autocorrelation but cannot infer the concentration of the population. Often it is necessary to identify whether the cluster is of high value population or of low value population indicating the severity of the variable under consideration. If high values, e.g. neighbourhoods with high crime rates or clusters with high income group, are to be identified or if clusters with low crime rate or low income group are to be identified in an area, then Moran's I and Geary's C cannot infer from the data population. Clusters with high value of the variable are often called hot spots. Similarly if low values cluster together these are called cold spots. These situations cannot be distinguished by the I and C ratios.

To identify and distinguish between the hot and cold spots in a spatially dispersed population the General G statistic has been developed which distinguishes between hot spots and cold spots. General G identifies the spatial concentrations of the population and is given by equation 9.12.

$$G(d) = \frac{\sum i \sum j W_{ij}(d) x_i x_j}{\sum i \sum j x_i x_j} \qquad (9.12)$$

where d is the neighbourhood distance between the observations x_i and x_j. The weight matrix w_{ij} has value 1 or 0 depending whether observation x_j is within distance d of x_i. The value of d the distance bound is generally judiciously choosen by the spread of the observations.

- The value of G is relatively large if high values cluster together.

- The value of G is relatively low if low values cluster together.

- The General G statistic is interpreted relative to its expected value (value for which there is no spatial association).

- Larger value of G or value larger than expected implies potential hotspot.

- Smaller value of G or value smaller than expected implies potential coldspot.

The G test statistic is used to see if the difference is sufficient to be statistically significant. Calculation of G must begin by identifying a neighbourhood distance within which a cluster is expected to occur.

- For the General G, the terms in the numerator (top) are calculated within a distance bound (d), and are then expressed relative to totals for the entire region under study.

- As with all of these measures, if adjacent x terms are both large with the same sign (indicating positive spatial association), the numerator (top) will be large.

- If they are both large with different signs (indicating negative spatial association), the numerator (top) will again be large, but negative.

9.7 Point Pattern Analysis in GIS

Most of the spatial data are modeled and processed by GIS in the form of points, lines and polygons representing discrete and continuous spatial objects. A dwelling area can be modeled in GIS as a set of points in a vector map. Therefore, to study the density and other factors it is necessary to study the pattern of points, and so point pattern analysis, pattern analysis, etc. are parts of inferential statistics.

How the point pattern deviates from a random pattern can be studied by computing the possible clustering among the point population. Sometimes it is better understood by applying a uniform grid to the sample area and analyzing the point pattern. This study is known as quadrant analysis. Other forms of inferential statistics are nearest neighbour analysis and spatial autocorrelation.

9.8 Applications of Spatial Statistical Methods

Spatial statistics finds its applications in many areas of GIS and statistical geography [8]. In research where spatial data are sampled across a region to study the spatio-temporal behavior of the region, spatio-statistical techniques are employed for deduction of effective inference. Following are some of the applications of spatial statistics.

1. To describe and summarize spatial data

2. To make generalizations concerning complex spatial patterns

3. To estimate the probability of outcomes for an event at a given location e.g. earthquake, tsunami etc.

4. To use the surveyed sample of spatial data to predict and infer the characteristics for the larger set of geographic data (population)

5. To determine if the magnitude or frequency of some phenomenon differ from one location to another

6. To compare and learn the spatial pattern of simulated data with the actual spatial pattern

9.9 Summary

This chapter starts with the definition of statistics. The definition is further extended to spatial statistics. The role of statistical methods in GIS is discussed. The difference between the univariate data with that of bivariate data is precursor to the understanding of the salient difference between the descriptive statistical methods and the inferential statistical methods. The two distinct types of statistical methods, the predictive statistics and the inferential statistical methods has been classified with the explanation of the differences between them. The different descriptive statistical methods such as the mean, mode, median, standard deviation, standard errors, range, skewness and kurtosis and their mathematical equations are then discussed. The second half of the chapter delves into inferential statistics. Inferential statistical methods such as Moran's Index, Geary's C and General G statistical index are explained with the equations and the associated parameters. How the spatial data is used to generate this index and how to interpret them to draw inference are explained with examples. The chapter ends with the explanation of spatial correlation and the spatial correlation coefficient.

10

An Introduction to Bathymetry

10.1 Introduction and Definition

Bathymetry is the science of measuring the depths of the oceans, seas, etc. and charting the shape and topography of the ocean floor. The name comes from Greek terms *bathus* meaning *deep*, and *metron* meaning *measure*. Hence bathymetry can be understood as the measurement of depth of deep sea or hydrographic body. In other words bathymetry is the study of underwater depth of lake or ocean beds. Bathymetric measurements can determine the topography of the ocean floor. Using bathymetric data one can study the sea floor, which is varied, complex, and ever-changing, containing plains, canyons, active and extinct volcanoes, mountain ranges, and hot springs. Some features, such as mid-ocean ridges where oceanic crust is constantly produced and subduction zones, also called deep-sea trenches, where it is constantly destroyed, are analyzed using the bathymetric data.

Bathymetric mapping involves the production of ocean and sea maps based upon bathymetric data. Bathymetric maps represent the ocean depth as a function of geographical coordinates in the same way topographic maps represent the altitude of Earth's surface at different geographic points. Bathymetry data can be used for modeling the sea floor surface digitally, and artificial illumination techniques are used to visualize the depths being portrayed. Paleobathymetry is the study of past underwater depths.

Bathymetric charts are typically produced to support safety of surface or subsurface navigation, and usually show sea floor relief or terrain as contour lines. Bathymetric contours are called depth contours or isobaths. Isobaths along with selected depths (soundings) provide crucial surface navigational information.

10.2 Bathymetric Techniques

For hundreds of years, the only way to measure ocean depth was the sounding line, a weighted rope or wire that was lowered overboard until it touched the ocean floor. Not only was this method time-consuming, it was inaccurate;

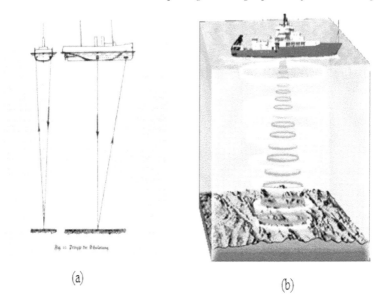

(a)

(b)

FIGURE 10.1
(a) Ray diagram of working sonar; (b) multi-beam sonar working principle

ship drift or water currents could drag the line off at an angle, which would exaggerate the depth reading. It was also difficult to tell when the sounding line had actually touched bottom.

In the twentieth century, sounding lines were entirely replaced by sonar systems. Sonar (**So**und **Na**vigation **R**anging), invented during World War 2 (1939–1945) measures depth by emitting a short pulse of high-frequency sound and measuring the time until an echo is heard. The data collected using sonar made it possible to prepare the complete bathymetric charts of the world's oceans. For the first time, scientists knew what 70% of Earth's surface really looked like. Radar, which produces images by bouncing radio waves rather than sound waves off distant objects, cannot be used for bathymetry because water absorbs radio waves.

Many sonar techniques have been developed for bathymetry. When high-resolution images are desired, an echo sounder (sonar) is mounted beneath or over the side of a boat, pinging a beam of sound downward at the sea floor. Remote sensing LIDAR or LADAR systems (side scan sonar see Figure 10.1(a)) are also used to collect high resolution bathymetric data. The amount of time it takes for the sound or light to travel through the water, bounce off the sea floor, and return to the sounder tells the equipment what the distance to the sea floor is.

Today, multi-beam echo sounders (MBES) are typically used, which use hundreds of very narrow adjacent beams arranged in a fan-like swath of typically 90 to 170 degrees across (See Figure 10.1b). The tightly packed array of narrow individual beams provides very high angular resolution and accuracy. In general a wide swath, which is depth dependent, allows a boat to map more sea floor in less time than a single-beam echo sounder by making fewer passes. The beams update many times per second (typically 0.1-50 Hz depending on water depth), allowing faster boat speed while maintaining 100% coverage of the sea floor. Attitude sensors allow for the correction of the boat's roll, pitch and yaw on the ocean surface, and a gyrocompass provides accurate heading information to correct for vessel yaw. (Most modern MBES systems use an integrated motion-sensor and position system that measures yaw as well as the other dynamics and position.) The global positioning system (or other global navigation satellite system (GNSS)) positions the soundings with respect to the surface of the Earth. Sound speed profiles (speed of sound in water as a function of depth) of the water column correct for refraction or 'ray-bending' of the sound waves owing to non-uniform water column characteristics such as temperature, conductivity, and pressure. A computer system processes all the data, correcting for all of the above factors as well as for the angle of each individual beam. The resulting sounding measurements are then processed either manually, semi-automatically or automatically (in limited circumstances) to produce a map of the area. As of 2010 a number of different outputs are generated, including a subset of the original measurements that satisfy some conditions (e.g. most representative likely soundings, shallowest in a region, etc.) or integrated digital terrain models (DTM) (e.g. a regular or irregular grid of points connected into a surface). Historically, selection of measurements was more common in hydrographic applications while DTM construction was used for engineering surveys, geology, flow modeling, etc. More recently, DTMs have become more acceptable in hydrographic practice.

10.3   Difference between Bathymetry and Topography

A **bathymetric chart** is the submerged equivalent of an above-water topographic map. Bathymetric charts are designed to present accurate, measurable descriptions and visual presentations of the submerged terrain.

In an ideal case, the joining of a bathymetric chart and topographic map of the same scale and projection of the same geographic area would be seamless. The only difference would be that the values begin increasing after crossing the zero at the designated sea level datum. Thus in topographic maps, mountains have the greatest values while in bathymetric charts,the greatest depths have the greatest values. Simply put, the bathymetric chart is intended to show the land if overlying waters were removed in exactly the same manner as the topographic map.

A bathymetric chart differs from a hydrographic chart in that accurate presentation of the underwater features is the goal, while safe navigation is the requirement for the hydrographic chart. A hydrographic chart will obscure the actual features to present a simplified version to aid mariners in avoiding underwater hazards.

Bathymetric surveys are a subset of the science of hydrography. They differ slightly from the surveys required to produce the product of hydrography in its more limited application and as conducted by the national and international agencies tasked with producing charts and publications for safe navigation. That chart product is more accurately termed a navigation or hydrographic chart with a strong bias toward the presentation of essential safety information. Bathymetric data is used for navigation in deep sea where safety to life is in question. Therefore, a standard has been evolved by the international maritime community known as Safety of Life at Sea (SOLACE). Bathymetry data need to comply to SOLACE inorder to be used for navigation in sea.

Bathymetric surveys and charts are closely tied to the science of oceanography, particularly marine geology, and underwater engineering or other specialized purposes.

10.4   Bathymetric Data Survey and Modeling

Unlike terrain surface, the sea is dynamic in nature and is shared by the international community for transportation, exploration and expedition. The sea is divided into two zones: (a) the exclusive economic zone (EEZ) of a country which extends to approximately twelve nautical miles from the shore and (b) international maritime zone which is beyond the EEZ. Therefore, the maritime survey rights of the EEZ lie with the respective country whereas the international maritime zone is surveyed by an organization known as the International Hydrographic Organisation (IHO). Therefore to share the survey data among the international community a common standard must be adhered to, which is designed and upgraded from time to time by the IHO. The standards such as S-52, S-57, S-63 and S-100 have been designed by the IHO for sharing of bathymetric data in the international maritime community and industry.

10.4.1   Bathymetric Data Models

Many IHO publications describing the bathymetric data model are available to the general public from the IHO website. Some important publications of IHO describing the bathymetric data standards and models are the *International Hydrographic Review, International Hydrographic Bulletin, Chart Specifications of the IHO*, and *The Hydrographic Dictionary*. The IHO has also published *Limits of Oceans and Seas*, which shows the boundaries between the oceans, as well as various international standards related to charting and hydrography, including S-57, the IHO transfer standard for digital hydrographic data.

10.4.1.1 S-57

S-57 is the IHO standard for the exchange of digital hydrographic data. S-57 is not the ENC (electronic navigation chart) product specification, but it is the generic framework standard for hydographic and related data. To date, it has been used almost exclusively for encoding ENCs; however there is a need for S-57 to support additional requirements. S-57 standards include:

- A general introduction with list of references and definitions.
- A theoretical data model on which the standard is based.
- The data structure or format that is used to implement the data model.
- General rules for encoding data in ISO/IEC 8211.

In addition to the main document, there are two appendixes in S-57.

- Appendix A is the object catalogue. It provides the official IHO-approved data schema that can be used within an exchange set to describe real-world entities.
- Appendix B contains the IHO-approved product specifications. These contain additional sets of rules for specific applications. Currently, the only product specification in S-57 is for an ENC.

There are several products based on the S-57 standards including:

- Additional Military Layers (AML)
- Marine Information Overlay (MIO)
- Inland ENC
- Port ENC

10.4.1.2 S-52

IHO standard S-52 provides colours and symbol specifications for ECDIS (Electronic Chart Display and Information System). Also it describes the presentation library for ECDIS (PresLib) which comprises a set of specifications, plus a symbol library, colour tables, look-up tables and symbolization rules. The PresLib links every object class and attribute of the ECDIS internal data base to the appropriate presentation of the ECDIS display. It provides details and procedures for implementing the display specifications.

10.4.1.3   S-63

S-63 is an IHO standard for encrypting and securing ENC data. The standard was adopted as the official IHO standard by the IHO member states in December 2002. The S-63 standard secures data by encrypting the basic transfer database using the Blowfish algorithm, SHA-1-hashing the data based on a random key and adding a CRC32 check. It also defines the use of DSA format signatures to authenticate the data originator.

10.4.1.4   S-100

S-100 came into use on 1 January 2010. It explains how the IHO will use and extend the ISO 1900 series of geographic standards for hydrographic, maritime and related issues. S-100 extends the scope of the existing S-57 hydrographic transfer standard. Unlike S-57, S-100 is inherently more flexible and makes provision for such things as the use of imagery and gridded data types, enhanced metadata and multiple encoding formats. It also provides a more flexible and dynamic maintenance regime via a dedicated on-line registry.

    S-100 provides the data framework for the development of the next generation of ENC products, as well as other related digital products required by the hydrographic, maritime and GIS communities. The key features of S-100 are:

- It provides interoperability with other ISO 19100 based profiles.
- It enables easier use of hydro data beyond HOs and ECDIS user's coastal zone mapping, security, inundation modeling.
- It provides plug-and-play updating of data, symbology and software enhancements.
- It supports imagery and gridded data, high-density bathymetry, and seafloor classification.
- It provides 3D and time-varying data (x,y,z and time).
- It supports marine GIS and web-based services.
- It supports gridded bathymetry.

10.5   Representation of Sea Depth and Sounding

Studying and visualizing the depths of the sea is like hovering in a balloon high above an unknown land which is hidden by clouds, because it is a peculiarity of oceanic research that direct observations of the abyss are impracticable. Instead of the complete picture that direct vision gives, we have to rely upon a patiently put together mosaic representation of the discoveries made from time to time by sinking instruments and appliances into the deep. Sea depths are collected as the sounding values obtained through sonar survey. The soundings are represented as nautical charts, electronic navigation charts, and paper navigation charts. These charts display the soundings and other hydrographic data according to the S-52 standards.

10.5.1 Nautical Chart

A nautical chart is a graphic representation of a maritime area and adjacent coastal regions. Depending on the scale of the chart, it may show depths of water and heights of land (topographic map), natural features of the sea bed, details of the coastline, navigational hazards, locations of natural and man-made aids to navigation, information on tides and currents, local details of the Earth's magnetic field, and man-made structures such as harbours, buildings, and bridges. Nautical charts are essential tools for marine navigation; many countries require vessels, especially commercial ships, to carry them. Nautical charting may take the form of charts printed on paper or computerised electronic navigation charts. Nautical charts are based on hydrographic surveys. As surveying is laborious and time-consuming, hydrographic data for many areas of sea may be dated and are not always reliable.

Nautical charts are issued by the national hydographic offices in many countries. These charts are considered 'official' in contrast to those made by commercial publishers. Many hydrographic offices provide regular, sometimes weekly, manual updates of their charts through their sales agents. Individual hydographic offices produce national chart series and international chart series. Coordinated by the IHO, the international chart series is a worldwide system of charts ('INT' chart series), which is being developed with the goal of unifying as many chart systems as possible. The nautical chart of New York Harbor is depicted in Figure 10.2.

10.5.2 Details on Nautical Chart

Conventional nautical charts are printed on large sheets of paper at a variety of scales. Electronic navigation charts, which use computer software and electronic databases to provide navigation information, can augment or in some cases replace paper charts, though many mariners carry paper charts as a backup in case the electronic charting system fails. Nautical charts must be labeled with aid to navigation or navigational information such as the pilotage information, depth information, tidal information etc.

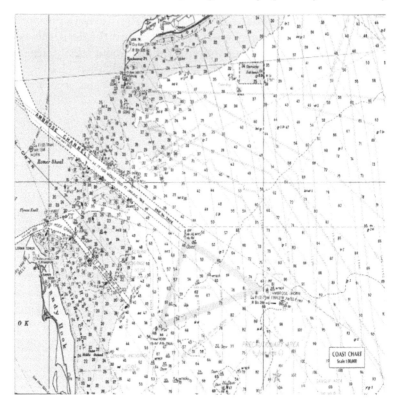

FIGURE 10.2
New York Harbor nautical chart

1. **Pilotage information:**
   The chart uses symbols to provide pilotage information about the
   nature and position of features useful to navigators, such as sea bed
   information, sea marks and landmarks. Some symbols describe the
   sea bed with information such as its depth, materials as well as possible hazards such as shipwrecks. Other symbols show the position
   and characteristics of buoys, lights, lighthouses, coastal and land
   features and structures that are useful for position fixing. Colours
   distinguish between man-made features, dry land, sea bed that dries
   with the tide and sea bed that is permanently underwater and indicate water depth.

2. **Depths:**
   Depths which have been measured are indicated by the numbers
   shown on the chart. Depths on charts published in most parts of
   the world use meters. Depth contour lines show the shape of underwater relief. Coloured areas of the sea emphasise shallow water

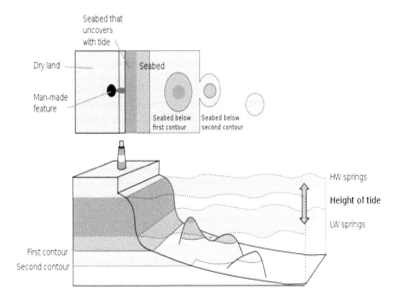

FIGURE 10.3
Chart colours and representation

and dangerous underwater obstructions. Depths are measured from the chart datum, which is related to the local sea level. The chart datum varies according to the standard used by each National Hydrographic Office. In general, the move is towards using lowest astronomical tide (LAT), the lowest tide predicted in the full tidal cycle, but in non-tidal areas and some tidal areas Mean Sea Level (MSL) is used.

Heights are generally given using highest astronomical tide (HAT) or mean sea level. The use of HAT for heights, and LAT for depths, means that mariners can quickly look at the chart to ensure that they have sufficient clearance to pass any obstruction, without the need to do tidal calculations each time.

3. **Tidal information:**
Tidal races and other strong currents have special chart symbols. Tidal flow information may be shown on charts using tidal-diamonds, indicating the speed and bearing of the tidal flow during each hour of the tidal cycle.

10.6 Map Projection, Datum and Coordinate Systems Used in Bathymetry

The Mercator projection is almost universally used in nautical charts. There are, however, some exceptions for very large or small scales where projections such as the gnomonic projection may be used. Since the Mercator projection is conformal, that is, bearings in the chart are identical to the corresponding angles in nature, bearings may be measured from the chart to be used at sea or plotted on the chart from measurements taken at sea.

Positions of places shown on the chart can be measured from the longitude and latitude scales on the borders of the chart, relative to a map datum such as WGS 84.

A bearing is the angle between the line joining the two points of interest and the line from one of the points to the north, such as a ships course or a compass reading to a landmark. On nautical charts, the top of the chart is always true north, rather than magnetic north, towards which a magnetic compass points. Most charts include a compass rose depicting the variation between magnetic and true north.

10.7 Application of Bathymetry Used in Preparation of bENCs

Untill recently, we did not have electronic navigation chart (ENC) with embedded bathymetry information. The bathymetric ENC (bENC) concept was developed by a company named Seven Cs and was presented in the September 2005 issue of *Hydro International, Ports and Harbors Special*. The approach aims to integrate the latest hydrographic survey data into ENC-based navigation software (see Figure 10.4). According to this concept, high-density bathymetric data are kept in separate S-57 data sets. bENCs are considered a bathymetric complement. The idea is to use bENCs in conjunction with regular ENCs (official or non-official) rather than to replace them, i.e. it does not involve any changes in topography or in the existing chart information, only in bathymetry. There was no need to introduce new S-57 object classes when the bENC concept was developed. bENCs make use of the same topology model as standard ENCs.

The integration of high-density bathymetry to ENCs has become a major issue. For this purpose, after acquiring, cleaning and modeling the hydrographic survey data, contours at meter (or sub-meter) intervals are generated. The contouring results are then converted into S-57 (the IHO exchange format

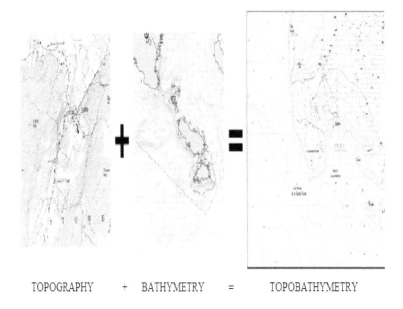

TOPOGRAPHY + BATHYMETRY = TOPOBATHYMETRY

FIGURE 10.4
Topobathymetry production of bENC

for ENCs) and encoded accordingly. The next step is the definition and attribution of depth areas, i.e. areas enclosed by adjacent contour lines. The result is a sort of electronic fair-sheet containing bathymetric data in S-57 format. This fair-sheet could be used to update the bathymetry of the official ENCs covering the relevant area. Data suitable for the production of high-density bathymetry ENCs (bENCs) are collected by port and waterway authorities on a regular basis. However, the data are usually not made available in S-57 format.

10.8   Differences between ENC, SENC, and RENC

Different types of navigation products and charts have been produced using the bathymetric data. The prominent charts used during sea voyages are electronic navigation charts, system electronic navigation charts and regional electronic navigation charts.

10.8.1   ENC - Electronic Navigational Chart

The data base, standardised as to content, structure and format, is issued for use with ECDIS. An ENC is equivalent to new editions of paper charts and may contain supplementary nautical information additional to that contained in the paper chart (e.g. sailing directions). The ENC is a subset of the ENC database developed by the national hydrographic authorities.

10.8.2   SENC - System Electronic Navigational Chart

The data base is transformed by ECDIS of the ENC for optimum use and updated by appropriate means. The SENC is the data base that is actually accessed for display generation and other navigational functions. The SENC contains the equivalent of the up-to-date paper chart. SENC distribution is a means of equipping an ECDIS with ENCs that is faster and generally more reliable than conventional distribution of ENCs in the S-57 format. SENC distribution involves converting the ENC data from the S-57 transmission format into the SENC format (a format specific to each ECDIS) in an office environment. This SENC file can then be sent to the ship and directly copied into the ECDIS.

10.8.3   RENC - Regional ENC Coordinating Center

The worldwide electronic navigational chart database (WEND) is the IHO network of hydrographic offices. It is the regional node responsible as issuing authority for official ENCs and official updates compiled of national ENC data.

10.9   Differences between a Map and a Chart

A chart, especially a nautical chart, has special unique characteristics including a very detailed and accurate representation of the coastline that takes into account varying tidal levels and water forms, critical to a navigator. A map emphasizes land forms, including the representation of relief, with shoreline represented as an approximate delineation usually at mean sea level. For a listing of the differences between the two, please see Table 10.1.

A chart is a working document used to plot courses for navigators to follow in order to transit a certain area. It takes into account special conditions required for one's vessel, such as draft, bottom clearance, wrecks and obstructions which can be hazardous. Way points are identified to indicate relative

CHART	MAP
A chart, especially a nautical chart, has special unique characteristics including a very detailed and accurate representation of the coastline, that takes into account varying tidal levels and water forms, critical to a navigator.	A map emphasizes land forms, including the representation of relief, with shoreline represented as an approximate delineation usually at mean sea level.
A chart is a working document used to plot courses for navigators to follow in order to transit a certain area. It takes into account special conditions required for one's vessel, such as draft, bottom clearance, wrecks and obstructions which can be hazardous. Way points are identified to indicate relative position and points at which specific maneuver such as changing courses must be performed.	A map is a static document that serves as a reference guide. It cannot be used to plot a course. It provides a predetermined course, usually a road, path, etc., to be followed. Special consideration for the type of vehicle is rarely present. Maps provide predetermined points-road intersections-to allow one a choice to change to another predetermined direction.
Charts provide detailed information about the area beneath the water surface normally not visible to the naked eye, which is critical for safe and efficient navigation.	Maps merely indicate a surface path providing no information of the condition of the road. A map will not provide information about whether the road is under repair (except when it is a new road) or how many pot holes or other obstructions it may contain. However the driver is able to make a visual assessment of such conditions.
NOAA Nautical Chart of Cape Henry, VA Chesapeake Bay.	USGS Topographic Map of Cape Henry, VA Chesapeake Bay.

TABLE 10.1
Differences between a Chart and a Map

position and points at which specific maneuvers such as changing courses must be performed.

A map is a static document that serves as a reference guide. A map cannot be used to plot a course. Rather it provides a predetermined course, usually a road, path, etc., to be followed. Special consideration for the type of vehicle

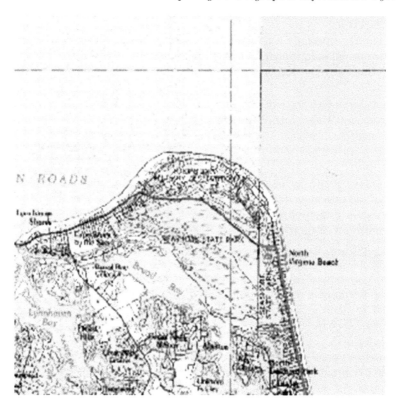

FIGURE 10.5
Example of a map

is rarely present. Maps provide predetermined points, road intersections etc. to allow one a choice to change to another predetermined direction.

Charts provide detailed information on the area beneath the water surface, normally not visible to the naked eye, which can be very critical for safe and efficient navigation. Maps merely indicate surface paths and provide no information about the condition of the road. A map will not provide information on whether the road is under repair (except when it is a new road) or how many potholes or other obstructions it may contain. However, the driver is able to make a visual assessment of such conditions.

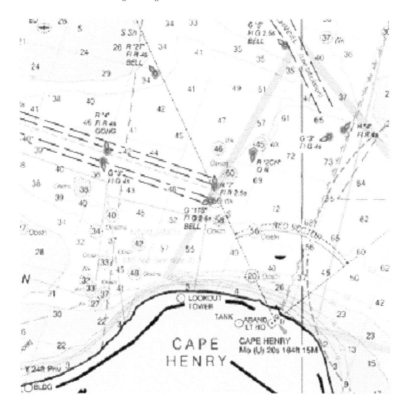

FIGURE 10.6
Example of a chart

10.10 Summary

This chapter introduces the science of bathymetry and its associated taxonomy to the reader. It gives a pedagogical definition of bathymetry followed by the different bathymetric survey techniques to measure the depth of sea floor. How the bathymetric data are modeled and stored in international standard formats such as S-52, S-57, S-63 and S-100 is discussed. How the sea depth is represented in a nautical chart is explained. The datum, map projection, and coordinate system used in bathymetry are discussed followed by different products of bathymetry such as the ENC, SENC and RENC. A classical difference between map and chart and different types of ENCs is given. Bathymetric concepts are integrated to the equivalent topographic concepts, explaining the spatial characteristics of the bathymetric survey data.

11

Spatial Analysis of Bathymetric Data and Sea GIS

The results of bathymetric or hydrographic surveys are produced in the form of nautical charts, sailing charts, coastal charts, harbor charts, electronic navigation charts, raster charts, etc. These charts in conjunction with computer aided digital display and sensors make excellent navigation equipments such as ECDIC (Electronic Chart Display and Information System), leveraging the power of bathymetric information and sensor information for safe navigation in deep sea. Therefore these systems are the best examples of manifestation of spatial information systems used for navigation, guidance and exploration of sea. Bathymetric information is also useful for effective planning of military operations at sea and for underwater exploration of natural resources.

One of the best examples of usage of bathymetric data is to identify a suitable site for resting or berthing of submarines. An undulating sea floor where the submarines can berth during operations is an important-spatio temporal decision. The captain of a submarine can using bathymetric data. The flat sea bed with the least slope and aspect in the vicinity of the current location needs to be computed and presented to the submarine commander in digital display format so as to arrive at a quick decision. This requires bathymetric sea bed surface generation and visualization in addition to the computation of slope and aspect.

Another example of a spatio-temporal decision to be made by the navigator of a ship is to find a suitable anchorage area near a harbor, or whether the ship can be safely guided to the harbor with appropriate jetty clearance, i.e. appropriate depth clearance of the sea floor from the hull of the ship, so that the ship does not ground. Depth contours give the general idea to a ship commander regarding the depth clearance while sailing in the deep sea. The scale of the depth information becomes very important when the ship is negotiating a harbor or approaching land.

Information such as the magnitude and direction of sea current, depth, surface and subsurface temperature, height of the waves and tide, land features which are aids to navigation such as location of lighthouses, beacons, sonobuoys, important landmarks, and islands are absolutely important for guiding a ship.

This chapter discusses the outputs and systems which are the outcome of bathymetry and hydrography. Systems such as ENC, SENC, ECDIC, and different forms of charts and their usefulness are discussed. The chapter starts with an answer to the question, what is a nautical chart? The contents of a nautical chart and the meta information associated with it are discussed followed by definitions of various charts with explanations of some important chart elements.

11.1 Difference between a Nautical Chart and an Electronic Chart

A nautical chart represents part of the spherical Earth on a plane surface. It shows water depth, the shoreline of adjacent land, prominent topographic features, aids to navigation, and other navigational information. It is a work area on which the navigator plots courses, ascertains positions, and views the relationship of the ship to the surrounding area. It assists the navigator in avoiding dangers and arriving safely at his destination. Originally hand-drawn on sheepskin, traditional nautical charts have for generations been printed on paper.

Electronic charts consisting of a digital database and a display system are in use and are replacing paper charts aboard many vessels. An electronic chart is not simply a digital version of a paper chart; it introduces a new navigation methodology with capabilities and limitations very different from paper charts. The electronic chart is the legal equivalent of the paper chart if it meets certain International Maritime Organization specifications.

Should a marine accident occur, the nautical chart in use at the time takes on legal significance. In cases of grounding, collision, and other accidents, charts become critical records for reconstructing the event and assigning liability. Charts used in reconstructing the incident can also have tremendous training value.

The different types of nautical charts depending upon the specific utilization and application of the ENC are sailing charts, general charts, coastal charts, and harbor charts. These are discussed below.

11.1.1 Sailing Charts

Sailing charts are the smallest scale charts. Sailing charts are used for planning, fixing position at sea and for plotting dead reckoning on a long voyage. The scale is generally 1:600,000 or smaller. The shoreline and topography are generalized and only offshore soundings, the principal navigational lights, outer buoys, and landmarks visible at considerable distances are shown in the chart.

11.1.2   General Charts

General charts are intended for coast-wise navigation outside of outlying reefs and shores. The scales range from about 1:150,000 to 1:600,000.

11.1.3   Coastal Charts

Coastal charts are intended for inshore coast-wise navigation, for entering or leaving bays and harbors of considerable width, and for navigating large inland waterways. The scales range from about 1:50,000 to 1:150,000.

11.1.4   Harbour Charts

Harbor charts are intended for navigation and anchorage in harbors and small waterways. The scale is generally larger than 1:50,000 showing the soundings of the paths to the harbour. The shape of the anchorage area and hazards are clearly depicted on the chart.

11.2   Projection Used in ENC

Cartographers cannot transfer a sphere to a flat surface without distortion so they project the surface of a sphere onto a developable surface. A developable surface can be flattened to form a plane. This process is known as chart projection. If points on the surface of the sphere are projected from a single point, the projection is said to be perspective or geometric. With widespread use of electronic charts, it is important to remember the cartographic principles applied to prepare paper chart are also applied to depict them on a video screen or electronic display.

The majority of NOAA (National Oceanic and Atmospheric Administration) charts use Mercator projection. This a cylindrical projection upon a plane with the cylinder tangent along the equator. The Mercator is the most common projection used in maritime navigation, primarily because rhumb lines are plotted as straight lines.

11.2.1   Some Characteristics of a Mercator Projection

1. Both meridians and parallels are expanded at the same ratio with increased latitude.

2. Expansion is the same in all directions, and angles are shown correctly (conformal).

3. Rhumb lines appear as straight lines, the directions of which can be measured on a chart.

4. Distances can be measured directly for accuracy.

11.2.2   Scale of ENC

The scale of a chart is the ratio of a given distance on the chart to the actual distance which it represents on the Earth. It may be expressed in various ways. The most common are:

1. A simple ratio or fraction, known as the representative fraction. For example, 1:80,000 or 1/80,000 means that one unit (such as a meter) on the chart represents 80,000 of the same unit on the surface of the Earth. This is sometimes called the natural or fractional scale.

2. A statement that a given distance on the Earth equals a given measure on the chart, or vice versa. For example, 30 miles to the inch means that 1 inch on the chart represents 30 miles of the Earth's surface. Similarly, 2 inches to a mile indicates that 2 inches on the chart represent 1 mile on the Earth. This is sometimes called the numerical scale.

3. A line or bar, called a graphic scale, may be drawn at a convenient place on the chart and subdivided into nautical miles, meters, etc.

All charts vary somewhat in scale from point to point, and in some projections the scale is not the same in all directions about a single point. A single subdivided line or bar for use over an entire chart is shown only when the chart is of such scale and projection that the scale varies a negligible amount over the chart, usually one of about 1:75,000 or larger. Since 1 minute of latitude is very nearly equal to 1 nautical mile, the latitude scale serves as an approximate graphic scale.

On most nautical charts the east and west borders are subdivided to facilitate distance measurements. On a Mercator chart the scale varies with the latitude. This is noticeable on a chart covering a relatively large distance in a north-south direction. On such a chart the border scale near the latitude in question should be used for measuring distances.

Of the various methods of indicating scale, the graphical method is normally available in some form on the chart. In addition, the scale is customarily stated on charts on which the scale does not change appreciably over the chart.

The ways of expressing the scale of a chart are easily interchangeable. For instance, in a nautical mile there are about 72,913.39 inches. If the natural scale of a chart is 1:80,000, one inch of the chart represents 80,000 inches of the Earth, or a little more than a mile. To find the exact amount, divide the scale by the number of inches in a mile, or $80,000/72,913.39 = 1.097$. Thus, a scale of 1:80,000 is the same as a scale of 1.097 (or approximately 1.1) miles to an inch.

Stated another way, 72,913.39/80,000 = 0.911 (approximately 0.9) inch to a mile. Similarly, if the scale is 60 nautical miles to an inch, the representative fraction is 1:(60 × 72,913.39) = 1:4,374,803.

A chart covering a relatively large area is called a small scale chart and one covering a relatively small area is called a large scale chart. Since the terms are relative, there is no sharp division between the two. Thus, a chart of scale 1:100,000 is large scale when compared with a chart of 1:1,000,000 but small scale when compared with one of 1:25,000.

As scale decreases, the amount of detail which can be shown decreases also. Cartographers selectively decrease the detail in a process called generalization when producing small scale charts using large scale charts as sources. The amount of detail shown depends on several factors, among them the coverage of the area at larger scales and the intended use of the chart.

11.3   Elements in a Bathymetric Chart

The chart title block should be the first thing a navigator looks at when receiving a new edition chart. The title tells what area the chart covers. The chart's scale and projection appear below the title. The chart will give both vertical and horizontal datums and, if necessary, a datum conversion note.

All depths indicated on nautical charts are reckoned from a selected level of the water called the sounding datum (sometimes referred to as the reference plane). For most NOAA charts of the United States in coastal areas, the sounding datum is Mean Lower Low Water (MLLW). In the Great Lakes, the plane of reference is the International Great Lakes Datum (1985).

Depths shown on charts are the least depths to be expected under average conditions. Since the chart datum is generally a computed mean or average height at some state of the tide, the depth of water at any particular moment may be less than shown on the chart. For example, if the chart datum is MLLW, the depth of water at lower low water will be less than the charted depth as often as it is greater.

Charts show soundings in several ways. Numbers denote individual soundings. These numbers may be either vertical or slanting; both may be used on the same chart, distinguishing between data based upon different US and foreign surveys, different datum, or smaller scale charts. Large block letters at the top and bottom of the chart indicate the unit of measurement used for soundings.

Soundings in fathoms indicates soundings are in fathoms or fathoms and fractions. Soundings in fathoms and feet indicates the soundings are in both fathoms and feet. A similar convention is followed when the soundings are in meters or meters and tenths.

Soundings are supplemented by depth contours, lines connecting points of equal depth. These lines present a picture of the bottom. On some charts depth contours are shown in solid lines; the depth represented by each line is shown by numbers placed in breaks in the lines, as with land contours. Solid line depth contours are derived from intensively developed hydrographic surveys. A broken or indefinite contour is substituted for a solid depth contour whenever the reliability of the contour is questionable.

Depth contours are labeled with numerals in the unit of measurement of the soundings. A chart presenting a more detailed indication of the bottom configuration with fewer numerical soundings is useful when bottom contour navigating. Such a chart can be made only for areas that have undergone a detailed survey. Shoal areas often are given a blue tint. Charts designed to give maximum emphasis to the configuration of the bottom show depths beyond the 100-fathom curve over the entire chart by depth contours similar to the contours shown on land areas to indicate graduations in height. These are called bottom contour or bathymetric charts.

The side limits of dredged channels are indicated by broken lines. The project depth and the date of dredging, if known, are shown by a statement in or along the channel. The possibility of silting is always present. Local authorities should be consulted for the controlling depth. NOS charts frequently show controlling depths in a table, which is kept current by the Notice to Mariners.

The chart scale is generally too small to permit all soundings to be shown. In the selection of soundings, least depths are shown first. This conservative sounding pattern provides safety and ensures an uncluttered chart appearance. Steep changes in depth may be indicated by more dense soundings in the area.

The limits of shoal water indicated on the chart may be in error, and nearby areas of undetected shallow water may not be included on the chart. Given this possibility, areas where shoal water is known to exist should be avoided. If the navigator must enter an area containing shoals, he must exercise extreme caution in avoiding shallow areas which may have escaped detection. By constructing a safety range around known shoals and ensuring his vessel does not approach the shoal any closer than the safety range, the navigator can increase his chances of successfully navigating through shoal water. Constant use of the echo sounder is also important.

Abbreviations listed in a chart are used to indicate what substance forms the bottom. The meaning of these terms can be found in the Glossary of this volume. While in the past navigators might actually have navigated by knowing the bottom characteristics of certain local areas, today knowing the characteristic of the bottom is most important when anchoring.

The nautical chart conveys a wealth of information to the mariner. Some examples of the type of information are listed below.

1. Floating aids to navigation established and maintained by the US Coast Guard mark channels and other features such as wrecks and obstructions.

2. The US Army Corps of Engineers dredges channels so that deep draft vessels can transit into and out of ports. Mariners must know the position and depth of these channels.

3. Nautical charts delineate the location of anchorages for military, commercial, and recreational vessels.

4. NOAA shows official geographic names in conformance with the US Board of Geographic Names.

5. Fixed aids to navigation, such as lighthouses maintained by the US Coast Guard, help mariners navigate safely.

6. Mariners need to know bottom characteristics in order to determine where adequate holding grounds for anchoring are located.

7. Depths determined by NOAA surveys are critical to safe navigation.

8. Mariners must know where underwater hazards and obstructions are located. The chart shows the precise position and depth of water over the obstruction.

9. Most commercial ships entering a harbour need to know where pilotage areas are located. These areas are used for taking on and dropping off marine pilots.

10. Mariners need to know the position and depths of dangerous wrecks, so they can lay out a track to avoid these features.

11. Wire drag cleared depths show the safe navigation depth. This charting symbol indicates that there is at least 20 feet of depth available over the top of the obstruction located here.

11.4 Summary

The applications of bathymetry or sea-GIS for navigation in the sea. The outcome of bathymetry is different types charts are discussed, including general charts, sailing charts, coastal charts and harbour charts. The characteristics of ENC are discussed. The cartographic projection used for preparation of charts is discussed. Various ways to represent scale of the chart are discussed. Understanding of the content of an ENC which can be considered as map equivalent of land is discussed as well as elements of a bathymetric chart. The difference between the map and bathymetric chart are brought out to clarify these concepts for the spatial science community dealing the topographic and bathymetric survey and applications separately. The datum, coordinate system and cartographic projection used in preparation of bathymetric data in general and chart in particular are discussed.

12

Measurements and Analysis Using GIS

Calculating the location, height, area, perimeter, volume, slope, aspect, distance between two points, line of sight between two points on the Earths surface etc. are some of the most fundamental and common computations in GIS. Advanced GIS often computes the azimuth and elevation of celestial objects from the Earth's surface and the almanac data pertaining to every location on the Earth for a given date and time. Sometimes an application-specific GIS can compute path profile between source and destination, crest clearance of a projectile, elevation profile of a terrain cut and radio line of sight between source and destination. Generally a GIS toolkit should have some of these computing tools in various forms. In this chapter we discuss the numerical formula of a representative set of such tools. It is not exhaustive as the set is ever increasing, with more sophisticated spatial computations emerging day by day.

12.1 Location

Location of an object in space is one of the intrinsic properties defining its position. It is expressed with respect to a frame of reference or coordinate system. Therefore to have an abstract idea about the location of an object in space it is mandatory to define its location in terms of coordinates. Hence location of an object is dependent upon the choice of frame of reference. A coordinate system is meaningful only when it is attached to a spatial model or to a datum surface. Therefore, intrinsically the coordinate of an object is dependent on the following:

Geodetic datum

Reference frame

Time of observation

Therefore, location is relative and is a dynamic quantity. Because in GIS we are mostly dealing with digital spatial data which has already been surveyed, the coordinate system, datum and the time of survey is already there in the form of a metadata. If we are observing the location of an object in sensor space through sensors such as UAV, GPS, radar etc. the time component of the location comes into play prominently. Therefore, GIS can be thought of as a system of computing methods to quantify, measure and visualize locations of spatial objects in different datums, frames of reference, coordinate systems and at different times.

Measurement of location in terms of spatial coordinates (latitude, longitude, altitude), spherical coordinates and equivalent (easting, northing and altitude) and rectangular coordinate systems is important for referencing any object on a digital map. Similarly the coordinate for objects at sea can be described through the latitude, longitude, and depth. The position description of any object in space is given by latitude, longitude and up, the height here can be orthomorphic height or height computed from the geoid surface known as geoidal height. Also for most purposes height is measured from the reference surface of the Earth known as the vertical datum. The commonly used vertical datum is Mean Sea Level (MSL) in the case of the Earth's surface and mean high tide for sea surface or depth data. GIS is a platform which gives the computing method to quantify and visualize location of spatial objects in different frames of reference, datums, coordinate systems and at different epochs of time. Some of the popular location measures and their applications are given in Table 12.1.

Measures of Spatial Position and Their Popular Applications	Usage in GIS
Geographic coordinate Latitude, Longitude, Height (ϕ, λ, h)	Survey and understanding of locations on the Earth's surface, making of maps.
Latitude, Longitude, Depth $(\phi, \lambda, Depth)$	Bathymetry survey, marine navigation and for preparation of bathymetric charts.
Easting, Northing, Altitude (ENA) where Easting and Northing can be of 2, 4, 6 or 10 digit precision	For location of military objects in the battle field. For engagement of targets by artillery force.
Polar coordinate in the form of (r, θ, ϕ)	For location of origin of weapons such as missiles, locating targets in unknown locations in the battle field, for navigation etc.

TABLE 12.1
Spatial Location Measures and Their Applications

12.2 Distance Measure

The notion of distance has many connotations in GIS. The distance which gives the sense of proximity of one object with respect to another object plays a crucial role in many spatial decisions in our daily life. Distance measure is one of the important metric measures that correlates different spatial objects. It imparts a topological relationship among objects and gives an idea of how near or how far the object is from another object. Inter se distance between spatial objects can be measured in many ways and therefore has many definitions and measures. Some of the significant distance measures computed by GIS and used in different applications are spatial distance, planimetric distance, cumulative distance, geodesic distance, nautical distance, Manhattan distance etc. The formula for computing these distances differs. Also while computing the distance the outcome depends on different datum, coordinate systems and projection systems used. Distance measure is one of the important co-relating metric measure between spatial objects. The significant distance measures which are computed by GIS are discussed here.

12.2.1 Linear Distance

Distance between any two points on a plane can also be given by simple Euclidean geometry. For any two points in the space with Cartesian coordinates (X_1, Y_1, Z_1) and (X_2, Y_2, Z_2) the Euclidian distance is given by the formula:

$$\text{distance} = \sqrt{(X_2 - X_1)^2 + (Y_2 - Y_1)^2 + (Z_2 - Z_1)^2}$$

In geodesy, the Euclidean formula can be used to calculate distance between two very near points (so that the surface can be approximated to be a plane). But this fails when larger distance is to be measured, as the curvature of the surface of Earth is significant. This distance in the Cartesian coordinate system is also known as geometric distance. This formula is used repetitively to compute the cumulative distance by taking the segment distance along the digitized vector joining the two points and adding them for the entire digitized vector.

12.2.2 Geodetic Distance

Computing distance between two points on a map has many connotations. The distance can be the aerial shortest distance (crow-fly distance), shortest distance as a line joining the two points drawn on flat surface, or the shortest distance out of many paths among the paths joining the two points. One possible computation is the shortest distance between two points $P_1(\Phi_1, \lambda_1)$

and $P_2(\Phi_2, \lambda_2)$ on the Earth's surface modelled as a sphere, depicted in Figure 12.1(a).

To find a suitable mathematical formula to compute the shortest distance, slice the spherical Earth along the two points and the center of the spherical Earth. One can imagine that the slice is a circle with center same as the center of Earth and P_1 and P_2 lying on its circumference. The shortest distance will be the arc length of the circle passing through these two points. If the radius of the spherical Earth is R then the spherical distance D is given by equation

$$D = RCos^{-1}(Sin\Phi_1 Sin\Phi_2 + Cos\Phi_1 Cos\Phi_2 Cos(\lambda_1 - \lambda_2)) \qquad (12.1)$$

This distance is known as spherical distance because it is measured for a spherical model. If the sphere is replaced with a suitable datum surface then the value of R will depend upon the semi-major axis, semi-minor axis and eccentricity of the datum. Then the distance computation is known as the geodesic distance.

12.2.3   Manhattan Distance

Manhattan distance is a form of taxicab geometry, described by Hermann Minkowski, a Japanese scientist in the 19th century, in which the usual distance function or metric of Euclidean geometry is replaced by a new metric in which the distance between two points is the sum of the absolute differences of their coordinates. The Manhattan distance is the simple sum of the horizontal and vertical components, whereas the diagonal distance might be computed by applying the Pythagorean theorem. The name Manhattan distance was inspired from the grid layout of streets on the island of Manhattan.

The Manhattan distance, d, between two vectors, p, q in an n-dimensional real vector space with fixed Cartesian coordinate system, is the sum of the lengths of the projections of the line segment between the points onto the coordinate axes. Manhattan distance is invariant to translation and reflection of the coordinate system, but it varies for the rotation of the coordinate system. Manhattan distance is depicted in Figure 12.1(b). Mathematically it can be expressed by the formula:

$$d(p, q) = || \, p - q \, || = \sum_{i=1}^{n} | \, p_i - q_i \, | \qquad (12.2)$$

where $p = (p_1, p_2, p_3......, p_n)$ and $q = (q_1, q_2, q_3......, q_n)$

12.2.4   Haversine Formula

Given a unit sphere, a triangle on the surface of the sphere is defined by the great circles connecting three points u, v, and w on the sphere. If the lengths

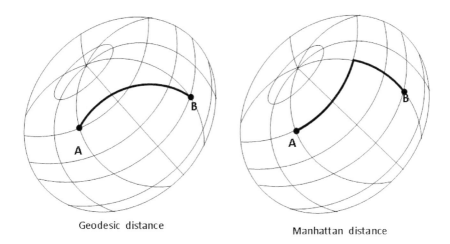

Geodesic distance

Manhattan distance

FIGURE 12.1
(a) Geodesic distance; (b) Manhattan distance

of these three sides are a (from u to v), b (from u to w), and c (from v to w), and the angle of the corner opposite c is C, then the law of haversines states:

$$haversin(c) = haversin(a - b) + sin(a)sin(b)haversin(C) \qquad (12.3)$$

Since this is a unit sphere, the lengths a, b, and c are equal to the angles in radians (using secant formula $\theta = \frac{l}{r}$ and for unit sphere i.e. $r = 1, \theta = l$) subtended by those sides from the center of the sphere. (For a non-unit sphere, they are the distances divided by the radius.)

The haversine formula is used to calculate the great-circle distance between two points; that is, the shortest distance over the Earth's surface distance between the points (ignoring the natural undulations).

12.2.4.1 Haversine Formula for Calculating Distance

To compute the distance between two points $P_1(\phi_1, \lambda_1)$ and $P_2(\phi_2, \lambda_2)$ haversine formula is used and is given by

$$haversin(\frac{d}{R}) = haversin(\Delta\phi) + cos(\phi_1)\,cos(\phi_2)haversin(\Delta\lambda) \qquad (12.4)$$

where R is mean radius of Earth (mean radius = 6371 km).

The haversine formula remains particularly relevant for numerical computation even at small distances unlike calculations based on the spherical law of cosines. The versed sine is $1 - \cos\theta$, and the half-versed-sine $\frac{1-\cos\theta}{2} = \sin^2(\frac{\theta}{2})$ as used above. It was published by Roger Sinnott in *Sky and Telescope* magazine in 1984 ("Virtues of the Haversine"), though known about for much longer by navigators. A marginal performance improvement can be obtained by factoring out the terms which get squared.

12.2.5   Vincenty's Formula

Thaddeus Vincenty devised a formula for calculating geodesic distances between a pair of points on the surface of Earth given by latitude and longitude. This uses an accurate ellipsoidal model of the Earth. While the mathematics is a bit difficult it is easy for programming. Vincenty's formula is accurate to within 0.5 mm on the ellipsoid being used. Calculations based on a spherical model, such as the haversine, are accurate to around 0.3% ,which is good enough for most purposes. The accuracy quoted by Vincenty applies to the theoretical ellipsoid being used, which differs from the real Earth geoid to varying degree.

12.3   Shortest Distance

Finding the shortest path from a source to destination is an important problem in many areas of application. Shortest path is a generic solution to many networks such as transport network, electrical network, communication network etc. For shortest path computation the underlined network is first modelled as a graph($G(v,e)$), where v is a set of vertices and e is a set of edges. There is a cost assigned to each edge in the network depending upon the modeling criteria. The cost associated with each edge differs for different networks e.g. in a road network the distance measures between two locations (vertices) is the edge strength where as in communication network bandwidth or the rate of data transfer is the strength of the edge. Similarly for an electrical network resistance is the strength of the edge. Hence given a network it is modelled to create edges having appropriate metric measures. Therefore shortest path computation is an optimisation problem to compute the path between the source vertices to the destination vertices which minimizes the cumulative cost of the edges. In the case of a transportation network the shortest between start and end vertices should be the shortest route in terms of ground distance. Yet another example of shortest path in a transportation network can be a path which minimizes the consumption of fuel or travelling time. Similarly in the case of a power transmission network the criteria of the shortest path can be to find a route which minimizes the power loss. In this section we investigate the shortest path computing method applicable to a homogeneous transport network which is modelled as a graph with the edge strength as ground distance.

Many algorithms have been proposed to compute shortest path of a graph. The prominent among them are:

Dijkstra's algorithm

Bellman-Ford algorithm

A-star algorithm

12.3.1 Dijkstra's Algorithm

Edsger Dijkstra (Turing Award, 1972) modified Ford's idea of shortest path (RAND, *Economics* of *transportation*, 1956) and published it, which today is well-known as Dijkstra's shortest path algorithm (published in 1959). It is just the simpler and faster (using relaxation and optimal selection) version of Ford's algorithm.

Let G(V, E) be a weighted graph with weight function w:E → R mapping edges to real valued weights. If e = (u, v), we write w(u, v) for w(e). The length of path p = $<v_0, v_1, v_2, .., v_k>$ is the sum of the weights of its consistent edges.

Therefore length(p) = $\sum_{i=1}^{k} w(v_{i-1}, v_i)$. The distance from u to v, denoted by (u, v) is the length of the path from u to v if there is a path and infinite otherwise.

Dijkstra's algorithm gives a solution to the class of problems known as **single source shortest path problems**. The problem goes as: given a directed graph (directed graphs are often denoted by digraph) with positive edge weights and a distinguished source vertex, s ∈ V, we will have to determine the distance and a shortest path from the source to every vertex in the digraph. Now the question is how to design an optimal and efficient algorithm for the problem. For this we will have to make some keen observations. Any sub-path of the shortest path must be a shortest path.

12.3.1.1 Intuition behind Dijkstra's Algorithm

1. Report the vertices in increasing order of their distance from source to vertex.

2. Construct the shortest path tree edge by edge, at each step adding one new edge corresponding to construction of shortest path to the current new vertex.

12.3.1.2 Idea of Dijkstra's Algorithm

1. Maintain an estimate d(v) of length $\delta(u, v)$ of the shortest path for each vertex v.

2. d[v] > δ(u, v) and d[v] equals the length of the known path i.e. d[v] = ∞ if we have no paths so far.

3. Initially d[s] = 0 and all the other d[v] values are set to ∞. The algorithm will process vertices one by one in some order.

Here processing a vertex u means finding new paths and updating d[u] ∀ v ∈ Adj[u] if necessary. The process by which an estimate is updated is called relaxation. The algorithm for relaxation (in pseudo code form) is given as:

```
Relax(u,v,w) {
if(d[u] + w(u, v) < d[v])
{
d[v] = d[u] + w(u, v);
pred[v] = u;
}
}
```

When all the vertices have been processed d[v] = δ(s, v) ∀ v.

Now the problem is to find the efficient order of processing the vertices. For this the greedy approach is being used. This approach is implemented using priority queue.

12.3.1.3   Pseudo Code for Dijkstra's Algorithm

```
Dijkstra(G,w,s) {
dist[s] ← 0; //Distance from s to v is initialized to 0.
for (∀ v ∈ V - {s})
do {
dist[v] ← ∞; // Set all the other distances to infinity.
s ← Φ; // S, the set of visited vertices is initially empty.
Q ← V; // Q, the set initially contains all the vertices.
}
while (Q != Φ) // While Q is not empty.
do {
u ← mindistance(Q, dist); // Select the element in Q with minimum distance.
s ← S ∪ {u}; // Add u to the visited vertices.
}
for (∀ v ∈ neighbours[u])
do {
Relax(u, v, w); // Calling of Relax function
}
return dist;
} end Dijkstra();
```

12.3.1.4 Analysis of the Time Complexity

The simplest implementation is to store the vertices in an array or linked list. But for sparse graphs or graphs with a lot of nodes but fewer edges, the efficient way is to store the graph by using binary heap or priority queue etc.

Many optimized alogorithms for computing shortest distance have been formulated. Some of the algorithms are based on greedy, dynamic programming or optimization approach. Prominent algorithms for computing the shortest path are the Bellman-Ford algorithm, the A-star algorithm etc.

12.3.2 Direction

Direction is the information pertaining to the relative position of one point with respect to another point. It is a relative quantity with respect to a frame of reference such as true north in a ECEF coordinate system. Direction also can be an absolute quantity measured in degree or radian with respect to some previously agreed frame of reference or object. It is often indicated by extending the index finger, or by the north arrow of the compass in a map. Mathematically direction is specified by a unit vector with respect to a set of axes defining the reference frame. In GIS direction is often measured by azimuth or bearing. The cardinal directions used are North (N), South (S), East (E) or West (W). For a better resolution intermediate cardinal directions such as North-East (NE), South-East (SE), South-West (SW) and North-West (NW) are used. A magnetic compass or a digital compass is the instrument that reads and measures the direction in a ECEF frame of reference that is fixed to the Earth. Direction is useful for navigation. Navigation is the field of study dealing with the process of monitoring and controlling the movement of a vehicle from the source to a destination. Usually one loses the sense of direction when navigating in the desert, sea, air or space. Therefore finding the current position of a craft, ship or self and orienting towards the destination through a predefined path calls for continuous monitoring of the current position and direction with respect to the destination. It is essentially the computation of position and direction. GIS with its digital maps assisted with a GPS and digital compass is an ideal system for navigation.

12.3.2.1 Azimuth

An azimuth is one way to define the direction from point to point on the ellipsoidal model of the Earth, such as Cartesian datum. Azimuth can be either measured clockwise from north through a full 360° or measured +180° in a clockwise direction from north and −180° in a counter clockwise direction from north.

On some Cartesian datum, an azimuth is called a grid azimuth, referring to the rectangular grid on which a Cartesian system is built. Grid azimuths are defined by a horizontal angle measured clockwise from north.

12.3.2.2   Bearings

Bearings is another method of describing directions. It is always an acute angle measured from 0° from at either north or south through 90° to either the west or the east. They are measured both clockwise and counter clockwise. They are expressed from 0° to 90° from north in two of the four quadrants the northeast, 1, and northwest, 4. Bearings are also expressed from 0° to 90° from south in the two remaining quadrants, the southeast, 2, and southwest, 3.

In other words bearing uses four quadrants of 90° each. A bearing of N 45° 15' 35" E is an angle measured in a clockwise direction 45° 15' 35" from north toward the east. A bearing of N 21° 44' 52" W is an angle measured in a counter clockwise direction 21° 44' 52" toward west from north. The same ideas work for southwest bearings measured clockwise from south and southeast bearings measured counter clockwise from south.

Azimuths and bearings are indispensable to locate a spatial object in geographic space. They can be derived from coordinates with an inverse calculation. If the coordinates of two points are geodetic, then the azimuth or bearing derived from them is also geodetic. If the coordinates from which a direction is calculated are grid coordinates, the resulting azimuth will be a grid azimuth, and the resulting bearing will be a grid bearing. Both bearings and azimuths in a Cartesian system assume the direction to north is always parallel with the y-axis, which is the north-south axis. On a Cartesian datum, there is no consideration for convergence of meridional, or north-south, lines.

12.3.2.3   North, Magnetic North and Grid North

The reference for directions is north. There are four categories of north used in GIS for different applications: (a) geodetic north, (b) astronomic north, (c) grid north and (d) magnetic north. Geodetic north differs from astronomic, which differs from grid north, which differs from magnetic north. The differences between the geodetic azimuths and astronomic azimuths are a few seconds of arc from a given point. Variations between these two are small compared to those found with grid azimuths and magnetic azimuths. For example, there is usually a difference of several degrees between geodetic north and magnetic north.

Magnetic north is used throughout the world as the basis for magnetic directions in both the northern and the southern hemispheres, but it will not hold still. The position of the magnetic North Pole is somewhere around 79° N latitude and 106° W longitude, a long way from the geographic North Pole, and it is moving. In fact, the magnetic North Pole has moved more than 600 miles since the early 19th century and it is still moving at a rate of about 15 miles per year, just a bit faster than it was previously.

The Earth's magnetic field is variable. For example, if the needle of a compass at a particular place points 15°, at the same place 20 years later that declination may have grown to 16° west of geodetic north. This kind of movement is called secular variation. Also known as declination, it is a change that occurs over long periods and is probably caused by convection in the material at the Earth's core.

Daily variation is probably due to the effect of the solar wind on the Earth's magnetic field. As the Earth rotates, a particular place alternately moves toward and away from the constant stream of ionized particles from the sun. Therefore, it is understandable that the daily variation swings from one side of the mean declination to the other over the course of a day. For example, if the mean declination at a place was 15° west of geodetic north, it might be 14.9° at 8 am, 15.0° at 10 am, 15.6° at 1 pm, and again 15.0° at shutdown. That magnitude of variation would be somewhat typical, but in high latitudes the daily variation can be as much as 9°.

The position of magnetic north is governed by natural forces, but grid north is entirely artificial. In Cartesian coordinate systems, whether known as state plane, Universal Transverse Mercator (UTM), a local assumed system, or any other system, the direction to north is established by choosing one meridian of longitude. Therefore, throughout the system, at all points, north is along a line parallel with that chosen meridian. This arrangement purposely ignores the fact that a meridian passes through each of the points and that all the meridians inevitably coverage with one another. Nevertheless, the directions to grid north and geodetic north would only agree at points on the one chosen meridian; at all other points there is an angular difference between them. East of the chosen meridian, which is frequently known as the central meridian, grid north is east of geodetic north. West of the central meridian grid north is west of geodetic north. Therefore it follows that east of the central meridian the grid azimuth of a line is smaller than its geodetic azimuth.

12.4   Area

Area measure expresses the expanse or vastness of a spatial object as perceived by an observer and projected to a frame of reference. In engineering practice, area measure plays a crucial role to quantify how much space is required to hold an object. Area is a fundamental parameter derived from terrain analysis, which is necessary for numerous decision-making processes. Generally distance and area are the fundamental terrain parameters associated with spatial data. They find many applications while supporting decision-making involving spatial data such as area analysis and cadastral applications. In most GIS, distance and area calculations are based on the vector data model, which makes a planar approximation of spatial data. In reality, the distance or area needs to be computed for an undulating surface. Given below is an approach to compute true area and distance from spatial vector data when the scale and projection are given. The spatial data is integrated with the elevation

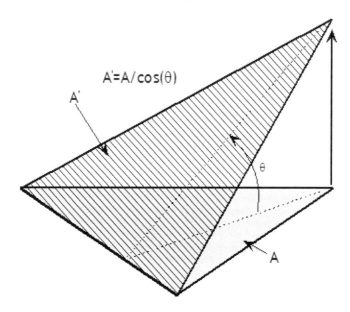

$A'=A/cos(\theta)$

FIGURE 12.2
Planimetric area of a triangle

model to compute slope and aspect of each and every point, thus preserving
the undulated property of the surface. The different area measures in GIS are
described here [2].

12.4.1   Planimetric Area

The planimetric area of a triangle is depicted in Figure 12.2 and is given by:

$$A_j = \frac{1}{2}(X_{2j}Y_{1j} + X_{3j}Y_{2j} + X_{1j}Y_{3j} - X_{1j}Y_{2j} - X_{2j}Y_{3j} - X_{3j}Y_{1j}) \quad (12.5)$$

Surface area of a triangle is given by:

$$A = \frac{1}{2}\sqrt{\begin{vmatrix} X_1 & Y_1 & 1 \\ X_2 & Y_2 & 1 \\ X_3 & Y_3 & 1 \end{vmatrix}^2 + \begin{vmatrix} Y_1 & Z_1 & 1 \\ Y_2 & Z_2 & 1 \\ Y_3 & Z_3 & 1 \end{vmatrix}^2 + \begin{vmatrix} Z_1 & X_1 & 1 \\ Z_2 & X_2 & 1 \\ Z_3 & X_3 & 1 \end{vmatrix}^2} \quad (12.6)$$

The other forms of area measures are orthogonal area measure and cumulative area measure. Cumulative area measure is an iterative or finite element
approach of the area measure given above.

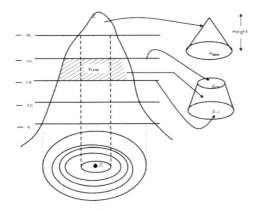

FIGURE 12.3
Computation of volume using contour data

12.5   Computation of Volume

Computing the volume of a given surface area and depth marked on a 2D map view or a 3D perspective view of terrain has many applications. Not many GIS available today can compute the volume of a chunk of Earth's surface from 2D spatial data stored as contours. Contours are iso-lines representing equal height on the ground. Generally they are concentric circular patches tagged with a height value. Computing volume from the contour data of a digital map involves the following steps.

The ground (defined by surface area and depth) for which volume has to be computed is cut into pieces along the contour planes. This results in a series of horizontal slabs as depicted in Figure 12.3. Each slab is treated as a prismoid with the height equals the contour interval and the end areas enclosed by the contour lines. Using the finite element principle the volume of the land mass is the sum total of volume due to these prismoids. Volume of the prismoid between the two contours C_i and C_j is given by

$$V_{C_i C_j} = CI \frac{A_{ci} + A_{cj}}{2} \qquad (12.7)$$

where CI is the contour interval measured from the map in meters.

Therefore the volume of the designated land mass is computed by adding the volume of the intermediate prismoid and the volume due to the tips of the contour. This is given by the formula

$$V = \frac{CI}{2}[A_{ci} + 2(A_{ci} + 2(A_{ci+1}) + A_{ci+2} +, ... + A_{cj-2} + A_{cj-1}) + A_{cj}] \quad (12.8)$$

12.6   Computation of Slope and Aspect

Slope can be defined as the degree of steepness or incline of a surface. Therefore slope is a continuous property of the terrain surface. Slope cannot be computed from the point clouds generated using LIDAR or some other means directly. To compute the slope one must first create either a raster grid, which is a gridded representation of pixels, or a TIN (Triangular Irregular Network) surface. Then the slope for a particular location in the surface is computed as the maximum rate of change of elevation between that location and its surroundings.

Slope can be expressed either in percentage or in degrees computed using the equations given below.

$$PercentSlope = \frac{Rise}{Run} x100 \tag{12.9}$$

$$DegreeSlope = arctan\frac{Rise}{Run} \tag{12.10}$$

Slope for a gridded DEM is computed using the finite difference methods. The height value of a 4×4 DEM grid locations is designated according to the cardinal direction from the center grid as depicted in Figure 12.4. in the Figure Z is the height at the middle cell and Z_e, Z_w, Z_s and Z_n are the height of East, West, South and North cells respectively. Similarly Znw, Zsw, Zne, Zse are the height of the Northwest, Southwest, Northeast and Southeast cells respectively. Let the inter se distance between two adjacent cells be D meters.

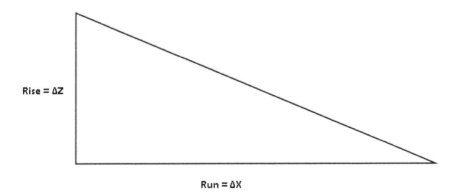

FIGURE 12.4
Slope computed as the ratio of rise over run in terrain surface

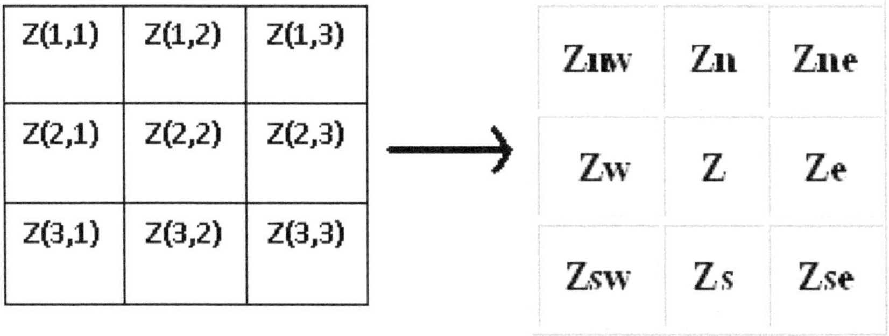

FIGURE 12.5
DEM grid with cardinal designator for the height

Then the slope can be computed using the slope along the X and Y direction is computed using the following difference equations.

$$\frac{\partial Z_X}{\partial X} = \frac{Z_e - Z_x}{2D} \tag{12.11}$$

$$\frac{\partial Z_Y}{\partial Y} = \frac{Z_e - Z_x}{2D} \tag{12.12}$$

If the slope of a middle cell in the grid is to be computed taking a 3×3 matrix of the DEM grid then the forward difference equations for computing the slope in cardinal direction are given by

$$\frac{\partial Z_x}{\partial X} = \frac{Z_{ne} + Z_e + Z_{se} - Z_{nw} - Z_w - Z_{sw}}{6D} \tag{12.13}$$

$$\frac{\partial Z_y}{\partial Y} = \frac{Z_{sw} + Z_e + Z_{se} - Z_{nw} - Z_n - Z_{ne}}{6D} \tag{12.14}$$

The slope at location (x, y) is computed using the above partial derivatives and is given by the equation

$$Slope = \sqrt{\left(\frac{\partial Z_x}{\partial X}\right)^2 + \left(\frac{\partial Z_y}{\partial Y}\right)^2} \tag{12.15}$$

Aspect is the orientation of slope, measured clockwise in degrees from 0 to 360, where 0 is facing north, 90 is east-facing, 180 is south-facing, and 270 is west-facing. The difference method of computing the slope and aspect at a grid location is given in the form of a difference equation 12.16.

$$Aspect = arctan\left(\frac{\frac{\partial Z}{\partial X}}{\frac{\partial Z}{\partial Y}}\right) \tag{12.16}$$

12.7   Curvature

The rate of change of slope i.e. the first order derivative of the slope or the second order derivative of DTM gives the curvature of the Earth's surface. Generally curvature describes the terrain surface in terms of how convex, concave or plane the surface is with respect to its surroundings. Curvature is computed using the second order partial derivatives as given by equation 12.17.

$$Curvature = \sqrt{\left(\frac{\partial^2 Z}{\partial X^2}\right)^2 + \left(\frac{\partial^2 Z}{\partial Y^2}\right)^2} \qquad (12.17)$$

The curvature of a particular point in the grid of DEM can also be computed using the Hessian function as given in Chapter 6 and its subsequent interpretation with respect to the physical property of the point on the Earth's surface.

The slope, aspect and curvature of a particular patch of Earth's surface are used to determine the patterns of flow of water, the flow acceleration, terrain change detection and land evaluation for different purposes etc.

12.8   Hill Shade Analysis

Hill shading is a technique used to visualize terrain as shaded relief. In this process the terrain surface is illuminated with a hypothetical light source. The illumination value for each raster cell is determined by its orientation to the light source, which is based on its slope and aspect. To simulate a natural landscape which has a good appreciation by the human cognition system it is advised to position the light source in the northwest which works best. Depending on the application, one can simulate the true position of the sun at a particular date and time of the year. Similarly, a shading pattern due to presence of the moon and its phase can be computed known as moon shaded relief map for appreciation of the visualization of the relief of the terrain surface.

12.9   Visibility Analysis

12.9.1   Line of Sight Analysis

Line of sight (LOS) analysis, also called 'view shed analysis', can be used to determine what can be seen from a particular location in the landscape. Con-

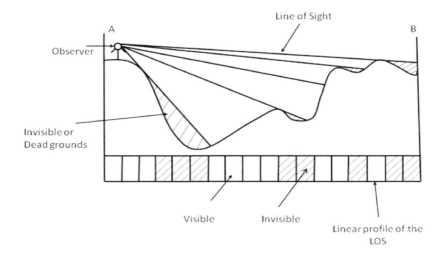

FIGURE 12.6
Line of sight between the observer and various points of the terrain

versely, the same analysis also determines from where within the surroundings that location can be seen. The first prerequisite for an LOS analysis is a three-dimensional surface model of the landscape in the form of DEM. For most applications, the most meaningful result would take vegetation, buildings, and other objects into account - those features not available in a bare-Earth digital elevation model (DEM). Above-ground features are usually included in a Digital Surface Model (DSM) which is created from LiDAR survey data. LOS computation has many variants depending upon potential usage in different applications. Computing the Optical LOS between the observer and the object is a popular function in every GIS. It has applications in surveillance in the battlefield by field observers of the border guard. In near-field observations the curvature of the Earth is not taken into consideration. Rather the local terrain undulation or the terrain profile plays a major role in deciding the LOS between the observer and the observent. This phenomena is depicted in Figure 12.6. For near-field observations using an optical aided instrument such as telescope, binocular or night vision device takes into account only the intervening obstacles such as high grounds or vegetation obstructing the LOS. Observing ships fading into the horizon or locating a lighthouse from a ship uses the principle of optical LOS taking into consideration the curvature of the Earth. Line of Sight Fan (LOS Fan) or 360 degree line of sight is computed for a complete visibility analysis around the observer. This is done to find out the dead zone or potential areas which cannot be observed manually.

Radio or RADAR Line of Sight (RLOS) is a computation performed to determine whether the communication line of sight is possible between the source transmitter of the radio waves and the destination receiver. RLOS is generally done over an extended area and beyond optical line of sight. Therefore bending of the radio waves due to atmosphere and multipath loss due to vegetation and intervening terrain is taken into consideration. While computing the RLOS, the Freshnel zone clearance is taken into consideration. The LOS concept is extended to compute the sphere of influence of the sensors such as sound sensor, heat sensor, microwave sensor etc. Sensor LOS is used to compute the sphere of influence of the sensors in the terrain covering the atmosphere too.

Some of the important uses of LOS analysis are for community planning and zoning, airport operations management, finding the ideal location of a camera or sensor for security coverage, finding the ideal location of the RADAR for battlefield coverage etc. Identifying a suitable location for cell phone tower placement is a direct application for LOS. While topography certainly has an impact on cell phone coverage, modeling cell phone signal propagation is in reality a much more complicated problem. An LOS analysis can be useful for planning cell phone tower placement, but to truly model cell phone coverage, more sophisticated models such as Freshnel's zone must be employed.

If H is the height of the observer in meters from the MSL, then the RLOS between the transmitter and listener D in kilometer. is given by equation

$$D = 3.57(\sqrt{KH_1}) \tag{12.18}$$

where K is an empirical constant adjusting for refraction of radio waves. Generally $K = 4/3$. As an offshoot of RLOS computation in GIS one can compute the height of the observer or antenna to be installed so as to view a particular target or range of the terrain at D kilometers from the transmitter. This leads to computing the LOS fan whereby the observer tries to see the entire terrain surrounding it by rotating his eye 360 degrees around its position. Here the visible areas are highlighted in green and invisible areas are highlighted in red. The distance between two communicating elements, e.g. a transmitter and antenna, can be computed using the formula given by:

$$D = 3.57(\sqrt{KH_1} + \sqrt{KH_2}) \tag{12.19}$$

where $K = \frac{4}{3}$ and H_1 and H_2 are the heights of the transmitter and antenna respectively.

In computing RLOS it is assumed that there is no intervening crest or physical obstruction between the transmitting and receiving elements. Therefore to compute the optical LOS the height profile of the intervening terrain between source and observer is computed for each point with a uniform range. The height profile is compared against the line joining the observer and the object point by point. If there is no height of terrain, that is more than the line joining the object and observer, then a LOS does not exist.

Imagine a sector of a circle which represents the Earth (like a slice of pie as depicted in Figure 12.7). At one end of the sector's arc erect a tower of height

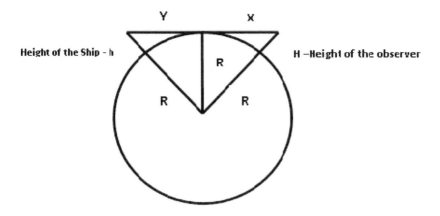

FIGURE 12.7
Line of sight between the observer and ship at sea

H and at the other end erect a tower of height h. The tops of these are the points between which we want to decide upon the longest LOS distance. Draw a line connecting the tops. We want to adjust the size of the sector so that this sight line is tangent to the circle at a point between the two bases. If the two towers were further apart, they could not see each other. As an offshoot of computation of LOS, a practical problem of a ship vanishing into the horizon at sea can be explained. The problem can be posed as follows. Find the LOS between an observer standing on a lighthouse which is at a height H meters from the MSL and the height of the ship is h meters. To compute the longest LOS consider Figure 12.7.

Consider the spherical Earth model with mean radius as R. Now draw a radius from the center of the Earth to the point of tangency between the LOS and the circle. Recall from basic geometry that the radius and the tangent line will be perpendicular to each other. Thus, we have two right angled triangles, one on either side of the radius to the tangent point. One of these triangles has hypotenuse $(H + R)$ and the other $(h + r)$ respectively. Therefore using the Pythagoras rule we obtain the following equations

$$(H + R)^2 = X^2 + R^2 \tag{12.20}$$
$$(h + R)^2 = Y^2 + R^2 \tag{12.21}$$

The maximum LOS distance is $(X+Y)$ beyond which the LOS will be a secant to the curvature of the Earth and the LOS will be obstructed and hence the observer will not be able to see the ship further. Solving the first equation above for X:

$$X^2 + R^2 = (H + R)^2 \tag{12.22}$$
$$X^2 + R^2 = H^2 + 2HR + R^2 \tag{12.23}$$
$$X^2 = H^2 + 2HR \tag{12.24}$$
$$X = \sqrt{H^2 + 2HR} \tag{12.25}$$

Following the same steps to solve the second equation for Y, we can then express $(X + Y)$ as:

$$X + Y = \sqrt{H^2 + 2HR} + \sqrt{h^2 + 2hR} \tag{12.26}$$

Therefore using the values of mean radius of Earth R, height of the mast of the ship h and the height of the observation point H one finds the maximum LOS between the fading ship and the observer.

12.10   Flood Inundation Analysis

Simulation of artificial flooding and analyzing its progress in time has many applications such as to simulate urban flooding patterns due to storm water, flooding due to hurricanes and high seas, to prepare emergency response teams for disaster and recovery planning etc. Listed below are few examples of usage of flood inundation simulation and analysis.

- Flood modeling is required to produce Digital Flood Insurance Rate Maps (DFIRM). The process of preparing DFIRMs involves spatial analysis of digital terrain model data, rainfall runoff or coastal storm surge models, hydrologic modeling, and hydraulic analysis to prepare a map which can classify the area into zones which are at high risk, moderate risk and no risk from flooding. This map is used by insurance agencies to decide the quantum and rate of insurance for different asserts.

- Flooding is used to cause potential delay or to stop an advancing contingent of armored vehicles such as tanks, armored carrier vehicles in a battle field. Field exercises are carried out to artificially flood potential areas where an armored contingent can penetrate. Exercises to study the progress of flooding in time and space and prepare flooding plans for strategic areas are carried out regularly by battle managers. These exercises help battle managers make decisions such as (a) suitable site for breaching (b) the dimension of the breach (c) amount of water required to cause effective flooding to cause the potential delay for the advancing armored contingent (d) monitoring the intensity and progress of flooding etc.

- Study of urban flooding and identification of low-lying areas susceptible to flooding due to storm water is a major concern for urban management bodies and government organizations engaged in disaster management. Also because of global warming the mean sea level is increasing, posing a threat to a vast population settled near the sea. Therefore flooding due to sea water is a real threat to human settlements near the sea. This unconventional flooding model need to be simulated and well understood to take precautionary measures before an actual disaster occurs.

To determine the potential depth of flooding, one must be able to predict how much water is in the watershed at any given time, how that amount of water changes over time during a storm event, and how the flow of water is impeded or obstructed by vegetation or man-made structures. Floodplain mapping comprises an entire engineering discipline in its own right in civil engineering. GIS are extensively useful in simulating, analyzing, and preparing topographic maps depicting flooding patterns.

The types of decision outputs and simulation results desired from flooding analysis are as follows.

- The ideal point of breach in an embankment so as to cause maximum flooding.
- The time required for flooding a particular zone or the time required for the flooding to affect a particular point.
- The amount of water required for the flooding to be effective.
- The size (width and depth) of the breach.
- Depth of flooding at a particular point in the flood zone.
- The safe area or safe zone in the flood zone where casualties can be evacuated to.

Simulation of a realistic flooding model requires the following spatio-temporal data as input.

- High quality DEM especially generated from a LiDAR-generated point cloud of height values.
- High resolution satellite image or aerial image of the area for realistic rendering of the flood zone.
- The water bodies and capacity their.
- Soil characteristics of the area.
- Weather and almanac information.

12.11   Overlay Analysis

Thematic composition of maps and creation of overlays depicting the spatio-temporal events occurring in discrete as well as continuous time-space are two important characteristics of GIS. The base map or the surveyed data is obtained after digitization of various themes. The overall map is known as the base map or composite base map. Users of a GIS can compose a digital map out of the base map according to the specific need of the application. In this, the surveyed digital map is seen as a composition of various thematic layers e.g. communication layer, road layer, rail layer, administrative zone, agriculture, etc. This layer can be further organized into sub-layers and sub-sub-layers depending upon the level of details of surveyed data. For example a road layer can have sub-layer as metal road, un-metal road, foot track etc. Further the metal road can be classified into a two lane bi-directional or one lane unidirectional road.

Therefore depending upon the application user can chose relevant themes out of the complete set of layers present in the back ground map. This composition of maps out of the composite map is known as 'application specific map composition' or 'thematic map composition' in GIS terminology.

Visualization and analysis of spatial data and events specific to an application requires plotting of the spatial data as an overlay with the base map in the background. This process is known as overlay analysis. In fact most of the usage and applications of GIS by different organizations is based on preparation of overlay based on the field survey data periodically or in a continuous time basis. Therefore overlay analysis is a spatio-temporal analysis. Based on the time interval overlay analysis can be divided into two categories, discrete time overlay analysis and continuous time overlay analysis, discussed below.

12.11.1   Discrete Time Overlay Analysis

Most common overlay analysis is a depiction of spatial situations or field survey data as an overlay in discrete time domain. In this process the data is collected through field survey in discrete time domain and plotted as an overlay with the base map in the background. Typical applications involve disaster management, city infrastructure planning, traffic management etc. For example the quantity of precipitation in different cities recorded in the past 24 hours recorded by weather stations across the country can be plotted as an overlay to analyze the effect of rain during the season. Speed and direction of the wind in the coastal towns during the day can be plotted to analyze the wind pattern in the form of a climo-graph. The percentage of citizens exercising their franchise across the country can be plotted as an overlay to analyze the political scenario. Most of the usage of overlay analysis using a GIS involves analysis of such data for understanding the spatio-temporal pattern in different applications. These spatial data are surveyed and compiled over a period of time and the statistics are depicted as an overlay with the base map in the background. There are many ways to depict the discrete samples for better visualization and analysis.

Overlays can be depicted in the form of surface density maps, point cloud maps, pi charts, bar graphs, colour coded maps etc.

The overlay analysis involving different spatial phenomena gives the user the correlation of one phenomena with respect to the other in space and time. This involves the set theoretic and algebraic operations such as union, intersection, set difference etc. performed on different overlays of the same area to identify the similarity, common and different events and their correlation. In fact there is an exhaustive set of overlay analysis known as map algebra.

12.11.2 Continuous Time Overlay Analysis

To analyze and visualize the spatial events occurring in continuous time domain the events are captured through sensors in discrete time space often known as the track data. These data are updated continuously with certain time intervals and depicted as a layer with the base map in the background. The overlay of tracks with a map background with continuous updates enables the user to analyze the movement pattern of the spatial object being tracked. A good example of these phenomena is the operation picture in the RADAR terminal tracking multiple flying objects. For this kind of overlay analysis the GIS is integrated with the sensor and the sensor input is piped to a common storage memory known as a buffer. The GIS continuously reads from the buffer and refreshes the track overlay. The typical application of these kinds of overlay analysis involves surveillance of the airspace, coastal zone or air-traffic management. The GIS in this situation is embedded in the data processing software of the sensor such as RADAR, SONAR, LiDAR etc.

12.12 Summary

This chapter discusses some of the measurement and analysis capability of GIS. The measurement and analysis are the outcome of computing algorithms applied on the spatial data. Spatial measures such as location, distance, direction, area, volume, slope, aspect, curvature etc. are discussed along with their variants. GIS is capable of analyzing spatial data and scenario many ways. Some of the analytical capabilities discussed are hill shed analysis to understand the undulation pattern of the terrain surface and line of sight analysis between the location of the observer and the location of the object. Flood inundation analysis and its applications are discussed. The generic overlay analysis capability of GIS to analyze and simulate spatio-temporal phenomena are discussed. GIS is capable of measuring and analyzing spatial objects and phenomena in multiple ways giving alterative perspective of the spatio-temporal phenomena.

13

Appendix A

13.1 Reference Ellipsoids

Ellipsoid	Semi-Major Axis	1/Flattening
Airy 1830	6377563.396	299.3249646
Modified Airy	6377340.189	299.3249646
Australian National	6378160	298.25
Bessel 1841 (Namibia)	6377483.865	299.1528128
Bessel 1841	6377397.155	299.1528128
Clarke 1866	6378206.4	294.9786982
Clarke 1880	6378249.145	293.465
Everest (India 1830)	6377276.345	300.8017
Everest (Sabah Sarawak)	6377298.556	300.8017
Everest (India 1956)	6377301.243	300.8017
Everest (Malaysia 1969)	6377295.664	300.8017
Everest (Malaysia and Sing)	6377304.063	300.8017
Everest (Pakistan)	6377309.613	300.8017
Modified Fischer 1960	6378155	298.3
Helmert 1906	6378200	298.3
Hough 1960	6378270	297
Indonesian 1974	6378160	298.274
International 1924	6378388	297
Krassovsky 1940	6378245	298.3
GRS 80	6378137	298.257222101
South American 1969	6378160	298.25
WGS 72	6378135	298.26
WGS 84	6378137	298.257223563

TABLE 13.1
Important Reference Parameters of Ellipsoids in Use

13.2   Geodetic Datum Transformation Parameters (Local to WGS-84)

- d = delta in meters

- e = error estimate in meters

- S = number of satellite measurement stations

Datum	Ellipsoid	dX	dY	dZ	Region of use	eX	eY	eZ	S
Adindan	Clarke 1880	-118	-14	218	Burkina Faso	25	25	25	1
Adindan	Clarke 1880	-134	-2	210	Cameroon	25	25	25	1
Adindan	Clarke 1880	-165	-11	206	Ethiopia	3	3	3	1
Adindan	Clarke 1880	-123	-20	220	Mali	25	25	25	1
Adindan	Clarke 1880	-166	-15	204	Mean for Ethiopia; Sudan	5	5	5	1
Adindan	Clarke 1880	-128	-18	224	Senegal	25	25	25	2
Adindan	Clarke 1880	-161	-14	205	Sudan	3	5	3	14

List of Geodetic Datum Transformation Parameters to (Local to WGS 84)

13.3 Additional Figures, Charts and Maps (please see color insert)

FIGURE 13.1
Satellite image of Chilka Lake in the state of Odisha in India depicting a land, sea and lake with its vector map draped on it

FIGURE 13.2
A contour map covering a portion of land and sea

FIGURE 13.3
Topobathymetry surface with vector data of topography and S-57 bathymetry
data of sea

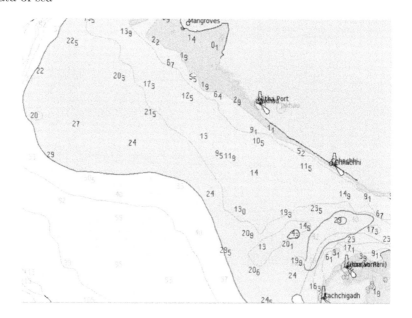

FIGURE 13.4
Topobathymetry surface depicting the sea contours and sounding measures of
the sea depth in fathoms

FIGURE 13.5
An instance of a flythrough visualization of a DEM draped with raster map

FIGURE 13.6
3D perspective visualization of an undulated terrain with sun shaded relief map draped on it

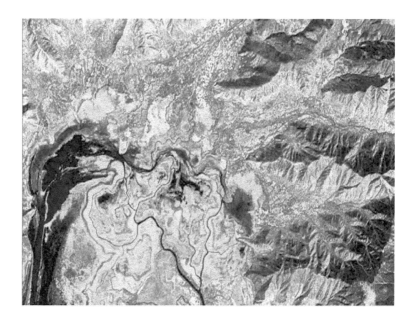

FIGURE 13.7
Colour-coded satellite image of an undulated terrain surface depicting relief

13.4 Line of Sight

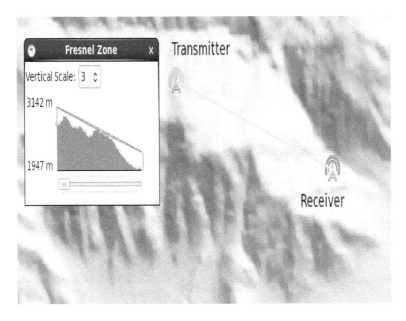

FIGURE 13.8
Computation of communication line of sight between transmitter and receiver
with the corresponding terrain profile along the LOS

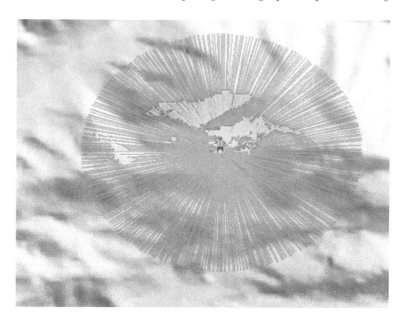

FIGURE 13.9
Computation of line-of-sight fan 360 degrees around the observer

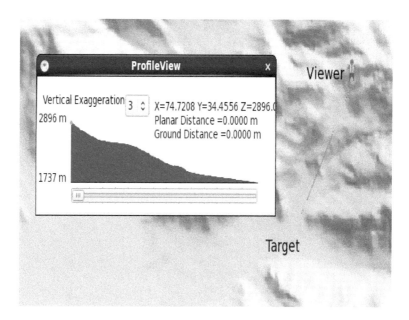

FIGURE 13.10
Line of sight between observer and the target the visible portion is depicted
in green and invisible in red

14

Appendix B

14.1 Definitions

14.1.1 Earth Sciences

Earth science, often referred as geological science in academia, is the branch of science concerned with determining the exact position of geographical points, the shape and size of the Earth. The branch of applied mathematics that deals with the measurement of the shape, area of large tracts of country, the exact position of geographical locations, the curvature, shape, and dimensions of the Earth is known as Earth science. It is the scientific study of the origin of the Earth along with its rocks, minerals, land forms, and life forms, and of the processes that have affected them over the course of the Earth's history, and the study of the structure of a specific region of the Earth, including its rocks, soils, mountains, fossils, and other features.

14.1.2 Geodesy

Geodesy is the scientific study of the size and shape of the Earth, its gravity field, and modeling of varying phenomena such as the motion of the magnetic poles and the tides; measurement of the dimension of Earth.

14.1.3 Geography

Geography is the study of the patterns and processes of human (built) and environmental (natural) landscapes, where landscapes comprise real (objective) and perceived (subjective) space. The study of the physical features of the Earth, its atmosphere, and of human activity as it affects and is affected by these. The science that deals with the Earth's physical structure and substance, its history, and the processes that act on it.

14.1.4 Bathymetry

The origin of the word bathymetry is traced to Greek literature. It is formed from the conjunction of two Greek words bathus (deep) and metron (measure). Bathymetry means the measurement of depth of water in oceans, seas, or lakes. The data generated by the measurements of depth of water through bathymetry are known as bathymetry data. Bathymetric charts also known as hydrographic charts are produced by plotting the bathymetric data and is typically used for surface and sub-surface navigation. Bathymetric charts also depict the seafloor relief as contour lines called depth contours or isobaths. Some selected depths (soundings) are also depicted in the bathymetric charts for safe navigation in the sea surface. The data used to make bathymetric maps today typically comes from an echosounder (sonar) mounted beneath or over the side of a boat, "pinging" a beam of sound downward at the seafloor or from remote sensing LIDAR or LADAR systems. The amount of time it takes for the sound or light to travel through the water, bounce off the seafloor, and return to the sounder informs the equipment of the distance to the seafloor. LIDAR/LADAR surveys are usually conducted by airborne systems.

14.1.5 Hypsometry

Hypsometry is a conjunction of the Greek words hupsos, which means "height" and, metron, which means "measure". Therefore hypsometry is the measurement of land elevations relative to the datum surface of Earth which is usually taken as the mean sea level. Hypsometry is the equivalent process to bathymetry underwater. A hypsometer is an instrument used in hypsometry, which estimates the elevation by boiling water. Water boils at different temperatures depending on the air pressure, and thus measures the altitude.

14.1.6 Hydrography

Hydrography refers to the mapping or charting of water's topographic features. It involves measuring the depths, tides, and currents of a water body and establishing the topography and morphology of seas, rivers, and lake beds. Normally and historically the purpose of charting a body of water is for the safety of shipping navigation. Such charting includes the positioning and identification of things such as wrecks, reefs, structures, navigational lights, marks and buoys and coastline characteristics.

14.1.7 Terrain

Terrain is a general term in physical geography referring to the lay of the land. It is usually expressed in terms of the elevation, slope, and orientation of terrain features.

14.1.8 Contour, Isoline, Isopleths

A contour line (also isoline, isopleth, or isarithm) is a function of two variables which is plotted as a curve along which the function has a constant value. In cartography, a contour line (often just called a "contour") joins points of equal elevation (height) above a given level, such as mean sea level. A contour map is a map which illustrates the relief of the terrain surface. The contour interval of a contour map is the difference in elevation between successive contour lines which defines the magnitude of the relative height between successive contours.

14.1.9 LIDAR

Lidar (also written LIDAR or LiDAR) is a remote sensing technology that measures distance by illuminating a target with a laser and analyzing the reflected light. The term "lidar" is the acronym for Light Detection and Ranging.

14.1.10 RADAR

RADAR an acronym for RAdio Detection And Ranging. It is an object detection system which uses radio waves to determine the range, altitude, direction, or speed of objects. It can be used to detect aircraft, ships, spacecraft, guided missiles, motor vehicles, weather formations, and terrain. The RADAR dish or antenna transmits pulses of radio waves or microwaves which bounce off any object in their path. The object returns a tiny part of the wave's energy to a dish or antenna which is usually located at the same site as the transmitter.

14.1.11 Remote Sensing

Remote sensing is the acquisition of information about an object or phenomenon without making physical contact with the object. In modern usage, the term generally refers to the use of aerial sensor technologies to detect and classify objects on Earth (both on the surface, and in the atmosphere and oceans) by means of propagated signals (e.g. electromagnetic radiation emitted from aircraft or satellites). There are two main types of remote sensing: passive remote sensing and active remote sensing. Passive sensors detect natural radiation that is emitted or reflected by the object or surrounding areas. Reflected sunlight is the most common source of radiation measured by passive sensors. Examples of passive remote sensors include film photography, infrared, charge-coupled devices, and radiometers. Active collection, on the other hand, emits energy in order to scan objects and areas whereupon a sensor then detects and measures the radiation that is reflected or backscattered from the target. RADAR and LiDAR are examples of active remote sensing where the time delay between emission and return is measured, establishing the location, speed and direction of an object.

14.1.12 Global Positioning System

Global Positioning System (GPS) is a space-based satellite navigation system that provides location and time information in all weather conditions, anywhere on or near the Earth where there is an unobstructed line of sight to four or more GPS satellites. The system provides critical capabilities to military, civil and commercial users around the world. It is maintained by the United States government and is freely accessible to anyone with a GPS receiver.

14.1.13   Principal Component Analysis

Principal component analysis (PCA) is a mathematical procedure that uses orthogonal transformation to convert a set of observations of possibly correlated variables into a set of values of linearly uncorrelated variables called principal components. The number of principal components is less than or equal to the number of original variables. This transformation is defined in such a way that the first principal component has the largest possible variance (that is, accounts for as much of the variability in the data as possible), and each succeeding component in turn has the highest variance possible under the constraint that it be orthogonal to (i.e., uncorrelated with) the preceding components. Principal components are guaranteed to be independent if the data set is jointly normally distributed. PCA is sensitive to the relative scaling of the original variables.

14.1.14   Affine Transformation

An affine transformation is any transformation that preserves collinearity and ratios of distances in the transformed image or map. That is to say all points lying on a line in the image still lie on a line in the image after transformation. Similarly the midpoint of a line segment remains the midpoint after transformation. In this sense, affine indicates a special class of projective transformations that do not move any objects from the affine space to the plane at infinity or conversely. An affine transformation is also called an affinity. While an affine transformation preserves proportions on lines, it does not necessarily preserve angles or lengths. Any triangle can be transformed into any other by an affine transformation, so all triangles are affine and, in this sense, affine is a generalization of congruent and similar. In general, an affine transformation is a composition of rotations, translations, dilations, and shears.

14.1.15   Image Registration

Image registration is the process of transforming different sets of data into one coordinate system. Data may be multiple photographs, data from different sensors, times, depths, or viewpoints. It is used in computer vision, medical imaging, military automatic target recognition, and compiling and analyzing images and data from satellites. Registration is necessary in order to be able to compare or integrate the data obtained from these different measurements.

14.1.16   Photogrammetry

Photogrammetry is the science of making measurements from photographs. The output of photogrammetry is typically a map, drawing, measurement, or a 3D model of some real-world object or scene. Many of the maps used are created with photogrammetry and photographs taken from aircraft.

14.1.17   Universal Transverse Mercator (UTM)

The Universal Transverse Mercator (UTM) geographic coordinate system is a 2D Cartesian coordinate system which references the locations on the surface of the Earth. It is a horizontal position representation, i.e. it is used to identify locations on the Earth independent of their vertical position, but differs from the traditional method of latitude and longitude in several respects. The UTM system is not a single map projection. The system instead divides the Earth into 60 zones, each a six-degree band of longitude, and uses a secant transverse Mercator projection in each zone.

15

Glossary of GIS Terms

Algorithm: A set of computing instructions for solving a specific problem having the following properties (a) finiteness (b) definiteness (c) input (d) processing (e) output. An algorithm transforms an input data to output information through the finite processing steps.

Attribute: Non-graphic or descriptive information describing the non-spatial properties associated with the spatial object modeled as a point, line, or area element in a GIS.

Autocorrelation: Also known as auto-covariance, is a statistical concepts expressing the degree to which the value of an attribute at spatially adjacent points varies with the distance or time separating the observations.

Base layer: A primary layer for spatial reference, upon which other layers are built. Examples of a base layer typically used are parcels, street centerlines or the survey map of the area.

Buffering: The creation of a zone of specific width around a point, line, or area. The buffer is a new polygon which is used in queries to determine which entities occur within or outside the defined area.

Computer-Aided Design (CAD): An automated system for the design, drafting and display of graphically oriented information.

Coordinate: A coordinate is a tuple specifying the location of objects with respect to the frame of reference under consideration. location in a 2D Cartesian coordinate system described by the tuple (x,y) and in 3D by the tuple (x,y,z).

Cadastral map: A map showing the precise boundaries and size of land parcels.

Cartography: The art and science of making of maps and charts.

Choropleth map: A map consisting of a series of single valued, uniform areas separated by abrupt boundaries set according to their attribute.

Classification: The process of assigning real objects to a group or set of surveyed data according to their attribute. In remote sensing image classification, the pixels are assigned to natural classes depending on the signature value recorded.

Composite map: A single map created by joining together several thematic layers that have been digitized separately.

Computer-Assisted Cartography (CAC): The use of computer hardware and specific software for making maps and charts.

Computing environment: The set of hardware and software facilities provided by a computer and its operating system with prime specifications of its processor, memory and operating system (OS).

Conceptual model: The abstraction, representation, and ordering of phenomena using the mind.

Contour: A line connecting points of equal elevation.

Convolution: The conversion of values from one grid to another which is different in terms of size or orientation.

Database: A logical collection of interrelated information managed and stored as a unit. A GIS database includes data about the spatial location and shape of geographic features recorded as points, lines, and polygons as well as their attributes.

Delaunay triangulation: A triangular irregular network of set of unique points in 2D plane having the properties (a) empty circumcircle; (b) the outer boundary of the triangulation is a convex hull.

Differentiable continuous surface: The representation of a continuously varying phenomenon using scalar or integer data so that the rate of change across and within the area may be derived.

Digital Elevation Model (DEM): A quantitative model of a part of the Earth's surface in digital form. Also known as digital terrain model (DTM). Digital elevation model is a grid of height values describing the terrain undulation.

Digitize: To encode map features as x,y coordinates in digital form. Lines are traced to define their shapes. This can be accomplished either manually or by use of a scanner.

Ellipsoid: Mathematical model for the shape of the Earth, taking into account of the flattening at the poles.

Exact interpolator: An interpolation method that predicts a value of an attribute at a sample point that is identical to the observed value.

Experimental variogram: An estimate of a semi-variogram based on sampling.

Extrapolation: The estimation of values of an attribute at unsampled points outside an area covered by existing measurements.

Finite difference modeling: A numerical modeling technique used with data held in regular grid form in which algebraic equations are used to solve changes in a variable at each location.

Finite element modeling: A numeric modeling technique used with data held in irregular grid (usually triangular) form in which algebraic equations are used to solve changes in a variable at each location.

Fourier analysis: A method of dissociating time series or spatial data into sets of sine and cosine waves.

Geocoding: The activity of defining the position of geographical objects relative to a standard reference grid. The process of identifying a location by one or more attributes from a base layer.

Grey scales: Levels of brightness (or darkness for displaying information on monochromic display devices). Generally the infinite pixel intensity of an image is scaled to value [0-255] for a gray scale image.

Grid: (a) A set of regularly spaced sample points. (b) A tessellation by squares. (c) In cartography, an exact set of reference lines over the Earth's surface. (d) In utility mapping, the distribution network of the utility resources, e.g. electricity or telephone lines.

Geographic Information System (GIS): (a) An organized collection of computer hardware, software, geographic data, and personnel designed to efficiently capture, store, update, manipulate, analyze, and display all forms of geographically referenced information. (b) A set of computer tools for collecting, storing, retrieving at will, transforming, and displaying spatial data from the real for a particular set of purposes.

Geographical primitives: The smallest units of spatial information: in vector form these are points, lines, and areas (polygons); in raster form they are pixels (2D) and voxels (3D).

Geo-reference: The referencing in space of the location of a point using a predefined coordinate system such a latitude and longitude or a national grid.

Global Positioning System (GPS): A satellite-based device that records x,y,z coordinates and other data. Ground locations are calculated by signals from satellites orbiting the Earth. GPS devices can be taken into the field to record data while walking, driving, or flying. A set of satellites in geostationary Earth orbits used to help determine geographic location anywhere on the Earth by means of portable electronic receivers.

Hidden line removal: A technique in 3D perspective graphics for suppressing the appearance of lines that ordinarily would be obscured from view.

Hypsometry: The measurement of the elevation of the Earth's surface with respect to sea level.

Indicator kriging: A kriging interpolation method which is non-linear and in which the original data are transformed from s continuous to a binary scale.

Inexact interpolator: Interpolation methods that provide estimates at data locations that are not necessarily the same as the original measurements.

Input: (noun) The data entered to a computer system; (verb) the process of entering data.

Interpolation: The estimation of values of an attribute at unsampled points from measurements made at surrounding sites.

Isoline: a line which joints locations of equal value.

Isopleth map: A map displaying the distribution of an attribute in terms of lines connecting points of equal value.

Kriging: Named after D.G. Krige, a set of interpolation techniques that use regionalized variable theory to incorporate information about the stochastic aspects of spatial variation when estimating interpolation weights.

Layer: A logical set of thematic data described and stored in a map library. Layers act as digital transparencies that can be laid atop one another for viewing or spatial analysis.

Linear interpolator: Describes a method whereby the weights assigned to different data points are computed using a linear function of distance between sets of data points and the point to be predicted.

Map: (a) A hand-drawn or printed document describing the spatial distribution of geographical features in terms of a recognizable and agreed symbolism. (b) A collection of digital information about a part of the Earth's surface.

Map projection: The basic system of coordinates used to describe the spatial distribution of elements in GIS.

Metadata: Data about the contents of the data. Information about a data set. Some key metadata are the source of the data; its creation date and format, its projection, scale, resolution, and accuracy, and its reliability with regard to some standard.

Nugget: In kriging and variogram modeling, that part of the variance of a regionalized variable that has no spatial component (variation due to measurement errors and short-range spatial variation at distances within the smallest inter-sample spacing).

Ordinary kriging: A method for interpolating data values from sample data using regionalized variable theory in which the prediction weights are derived from a field variogram model.

Ortho imagery: Aerial photographs that have been rectified to produce an accurate image of the Earth by removing tilt and relief displacements.

Orthophotos: A scale-correct photomap created by geometrically correcting aerial photographs or satellite images.

Output: The results of processing data in a GIS; maps, tables, screen images, tape files.

Overlay: (verb) The process of stacking digital representations of various spatial data on top of each other so that each position in the area covered can be analyzed in terms of these data. (noun) A data plane containing a related set of geographic data in digital form.

Point: A single (x,y) coordinate that represents a geographic feature too small to be displayed as a line or area at that scale.

Polygon: A figure that represents an area on a map generally encoded as a closed sequence of line segments. Polygons have attributes that describe the geographic feature they represent.

Photogrammetry: A series of technique for measuring position and altitude from aerial photographs or images using a stereoscope or stereoplotter.

Pixel: An abbreviation for picture element; smallest unit of information in a raster map such as DEM or scanned image.

Quadtree: A data structure for thematic information in a raster database that seeks to minimize data storage.

Raster data structure: A database containing all mapped spatial information in the form of regular grid cells.

Raster display: A device for displaying information in the form of pixels on a computer screen or VDU.

Raster map: A map encoded in the form of a regular array of cells known as pixels.

Resampling: A technique for transforming a raster image from one particular scale and projection to another.

Resolution: The smallest spacing between two displayed or processed elements; the smallest size of feature that can be mapped or sampled.

R-tree: A spatial indexing technique which groups entities according to their proximity by using minimum bounding rectangles. Hierarchies of rectangles may be established. When querying the database any search is directed to the rectangle and any subsequent lower-level ones which contain the item of interest.

Run-length codes: A compact method of storing data in raster databases which simplifies the grid on a row-by-row basis by coding the start and end values of contiguous cells for each class.

Sampling: The technique of obtaining a series of measurements to obtain a satisfactory representation of the real world phenomenon being studied.

Scale: The metric property of a map or an image which defines the ratio between a distance or area on a map and the corresponding distance or area on the ground.

Scanner: A device for converting images from maps, photographs, or from part of the real world into digital form. The scanning head is made up of a light or other energy source and a sensing device which records digital values of light reflected back from the surface.

Semivariogram: (a) Given two locations x and (x+h), a measure of one-half of the mean square differences (the semivariance) produced by assigning the value z(x +h) to the value z(x), where h (known as the lag) is the inter sample distance. (b) A graph of semivariance versus lag h.

Semivariogram model: One of a series of mathematical functions that are permitted for fitting the points on an experimental variogram (linear, spherical, exponential, Gaussian, etc.).

Sill: The maximum level of semivariance reached by a transitive semivariogram.

Simple kriging: An interpolation technique in which the prediction of values is based on a generalized linear regression under the assumption of second order stationarity and a known mean.

Simulation: Using the digital model of the landscape in a GIS for studying the possible outcome of various processes expressed in the form of mathematical models.

Spatial analysis: The process of modeling, examining, and interpreting model results. Spatial analysis is useful for evaluating suitability and capability, for estimating and predicting, and for interpreting and understanding.

Structured Query Language (SQL): A syntax for defining and manipulating data from a relational database. Developed by IBM in the 1970s, it has become an industry standard for query languages in most relational database management systems.

Tessellation: The process of dividing an area into smaller, contiguous tiles with no gaps in between them.

Thematic map: A map displaying selected kinds of information relating to specific themes, such a soil, land use, population density, suitability for arable crops, and so on. Many thematic maps are also choropleth maps, but when the attribute is modeled by a continuous field, representation by isolines or colour scales is more appropriate.

Thiessen polygons: A tessellation of the plane such that any given location is assigned to a tile according to the minimum distance between it and a single, previously sampled point. Also known as Dirichlet tessellation or Voronoi polygons.

Topographical map: A map showing the surface features of the Earth's surface (contours, roads, rivers, houses, etc.) in great accuracy and detail relative to the map scale used.

Topology: A term used to refer to the continuity of space and spatial properties, such as connectivity, that are unaffected by continuous distortion. In the representation of vector entities, connectivity is defined explicitly by a directed pointer between records describing things that are somehow linked in space (for example a junction between two roads). In regular and irregular tessellations of continuous surfaces (e.g. grids) the topological property of connectivity between different locations may only be implicitly defined by the spatial rate of change of attribute values over the grid. The topology (connectivity) of gridded surfaces can be revealed by computing first, second, or higher order derivatives of the surface

Transformation: The process of changing the scale, projection, or orientation of a map or an image.

Trend surface analysis: Methods for exploring the functional relationship between attributes and the geographical coordinates of the sample points.

Triangular Irregular Network (TIN): A vector data structure for representing geographical information that is modeled as a continuous field (usually elevation) which uses tessellated triangles.

Universal kriging: A simple kriging of the residuals of a regionalized variable after systematic variation has been modeled by a drift or trend surface.

Vector: (a) In Physics, a quantity having both magnitude and direction. (b) In GIS, the representation of spatial data by digitized points, lines, and polygons.

Vector data structure: A means of coding and storing point, line, and areal information in the form of units of data expressing magnitude, direction, and connectivity.

Viewshed: A visualization technique in which those parts of the landscape are visualized which can be seen from a particular point.

Voxels: Three-dimensional cubic units of space.

Weighted moving average: The value of an attribute computed for a given point as an average of the values at surrounding data points taking account of their distance or importance.

Bibliography

[1] E.H. Adelson, E.P. Simoncelli, and R. Hingorani. Orthogonal pyramid transforms for image coding. In *Proc SPIE Visual Communications and Image Processing II*, 845: pp. 50–58, Cambridge, MA, October 1987.

[2] M.R. Anstaett. Area calculations using Pick's theorem on freeman-encoded polygons in cartographic systems. *Proceedings of AUTO-CARTO 7, Seventh International Symposium on Computer-Assisted Cartography*: 11–21, 1985.

[3] F. Aurenhammer. Voronoi diagrams: A survey of fundamental geometric data structure. *ACM Computer Survey*, 23:345–405, 1991.

[4] T. Bernhardsen. *Geographical information systems: An introduction.* 2002.

[5] A. Bowyer. Computing dirichlet tessellation. *The Computer Journal*, 24(2):162–166, 1981.

[6] J.J. Brannan, D.A. Esplen, and M.F. Gray. *Geometry.* 2nd Ed., 2012.

[7] L.G. Brown. A survey of image registration techniques. *ACM Computing Surveys*, 24(4):989–1003, 1992.

[8] P.A. Burrough and A.R. McDonnell. *Principles of geographical information systems.* 1998.

[9] P.J. Burt, and E.H. Adelson. The Laplacian pyramid as a compact image code. *IEEE Transactions on Communications*, 31:532–540, 1983.

[10] J. Canny. A computational approach to edge detection. *IEEE Transactions on Pattern Analysis and Machine learning*, 8(6):679–698, 1986.

[11] J. Christopher. *Geographical information systems and computer cartography.*

[12] C. Clarke. *Getting started with geographic information systems.* 2nd Ed., 1997.

[13] T.H. Cormen, R.L. Leiserson, and C.E. Rivest. *Introduction to algorithms* 2nd Ed. *C. Stein, ed.*, MIT Press and McGraw-Hill, 2001.

[14] J. Dangermond. A review of digital data commonly available and some of the practical problems of entering them into a GIS. *Introductory Readings in Geographic Information Systems*, pp. 222–232, 1990.

[15] D. Fenna. *Cartographic science.* 2007.

[16] G. Dutton. Harvard papers on geographic information systems. *First Advanced study Symposium on Topological Data Structures for Geographic Information Systems*, Reading MA: Addison-Wesley, 1979.

[17] S. Fortune. A sweep line algorithm for Voronoi diagram. *Algorithmica*, 2(2):153–174, 1987.

[18] J.M. Gauch and S.M. Pizer. Multiresolution analysis of ridges and valleys in grey-scale images. *IEEE Transactions on Pattern Analysis and Machine Intelligence*, 15:635–646, 1993.

[19] G. Joseph. *Fundamentals of remote sensing.* 2nd Edn, 2005.

[20] M.F. Goodchild. Spatial autocorrelation. *Concepts and Techniques in Modern Geography*, 1991.

[21] M.F. Goodchild. Geographic information science. *International Journal of Geographic Information Systems*, 6:31–45, 1992.

[22] A.A. Goshtasby. *2-d and 3-d image registration for medical, remote sensing and industrial applications.* John Wiley, pp. 112–115, 2005.

[23] R. Green, P.J. Sibson. Computing Dirichlet tesselation in the plane. *The Computer Journal* (24), 1981.

[24] C. Harris and M.J. Stephens. Corner and edge detector a computational approach to edge detection. In *Computer Vision Conference*, pp. 147–152, 1988.

[25] I. Heywood, S. Cornelius, and S. Carver. *An introduction to geographical information systems.* 2002.

[26] J.R. Jensen. *Introductory digital image processing: A remote sensing perspective.* 1986.

[27] Y.J. Kaufman. The atmospheric effect on remote sensing and its correction, theory and applications of optical remote sensing. G. Asrar (ed.), John Wiley 1989.

[28] J.J. Koenderink and F. Dillen. Image space. In: *Pure and applied differential geometry.* Woestyne, ed., Shaker Verlag, Aken, Germany, pp. 149–157, 2008.

[29] J.J. Koenderink and A.J.V. Doorn. The structure of visual spaces. *Journal of Mathematical Imaging and Vision*, 31:171–187, 2008.

[30] J.J. Koenderink and S.C. Pont. Material properties for surface rendering. *International Journal for Computational Vision and Biomechanics*, 1(1):45–53, 2008.

[31] P.A. Longley, M. Goodchild, D.J. Maguire, and D.W. Rhind. *Geographic information systems*. Vol 1 and Vol 2:31–45, 1999.

[32] D. Lowe. Object recognition from local scale-invariant features. In *Proceedings of the International Conference on Computer Vision*, pp. 1150–1157, 1999.

[33] D. Lowe. Distinctive image features from scale-invariant keypoints. *International Journal of Computer Vision*, 60(2):91–110, 2004.

[34] J.B. Maintz and M.A. Viergever. A survey of medical image registration. *Medical Image Analysis*, 2(1):1–37, 1998.

[35] D.H. Maling. *Coordinate systems and map projections*. 2nd Ed., 1973.

[36] S.G. Mallat. A theory for multiresolution signal decomposition: The wavelet representation. *IEEE Transactions on Pattern Analysis and Machine Intelligence*, 11:674–693, 1989.

[37] P. Meer, E.S. Baugher, and A. Rosenfeld. Frequency domain analysis and synthesis of image pyramid generating kernels. *IEEE Transactions on Pattern Analysis and Machine Intelligence*, 9:512–522, 1987.

[38] N. Mirante and A. Weingarten. The radial sweep algorithm for constructing triangulated irregular network. *IEEE Computer Graphics and Applications*, Vol 2, 1982.

[39] C. Morino. Efficient 2-d geometric operations. *C/C++ User Journal*, 25–36, 1998.

[40] J. O'Rourke.. *Computational geometry using C*. 2nd Ed., 1998.

[41] N. Panigrahi. *Geographic information science*. 2008.

[42] N. Panigrahi, B.K. Mohan, and G. Athithan. Terrain modelling using dominant points. *Proceedings of Second International Advance Computing Conference*, IACC 2010, Patiala, India, pp. 34–37, 2010.

[43] N. Panigrahi, B.K. Mohan, and G. Athithan. Classification of changed pixels in satellite images using Gaussian and Hessian functions. *International Conference on VLSI, Communication and Instrumentation, ICVCI*, 2011.

[44] N. Panigrahi, B.K. Mohan, and G. Athithan. Differential geometric approach to change detection using remotely sensed images. *Journal of Advances in Information Technology*, 2(3):134–138, 2011.

[45] N. Panigrahi, B.K. Mohan, and G. Athithan. Pre-processing algorithm for rectification of geometric distortions in satellite images. *Defence Science Journal*, 61:174–179, 2011.

[46] R.J. Little, J.J. Peucker, T.K. Fowler, and D.M. Mark. Digital representation of three-dimensional surface by triangulated irregular network. *Technical Report No-10, US Office of Naval Research, Geography Programs*, 10, 1976.

[47] T.K. Peucker and N. Chrisman. Cartographic data structures. *American Cartographer*, 2:55–69, 1975.

[48] W.K. Pratt. *Digital image processing*, 2nd Ed. John Wiley, 1991.

[49] M.I. Preparata, F.P. Shamos. *Computational geometry: An introduction.* 2nd Ed., 1993.

[50] S. Ranganath. Image filtering using multiresolution representations. *IEEE Transactions on Pattern Analysis and Machine Intelligence*, 13:426–440, 1991.

[51] S. Rebay. An efficient unstructured mesh generation by means of delaunay triangulation and Bowyer-Watson algorithm. *Journal of Computational Physics*, 105:125–138, 1993.

[52] J.P. Snyder. *Map projections used by the USGS.* 2nd Ed., 1983.

[53] J.P. Snyder. The transverse and oblique cylindrical equal-area projection of the ellipsoid. *Annals of the Association of American Geographers*, 75:431–42, 1985.

[54] J.P. Snyder. *Flattening the Earth: Two thousand years of map projection.* 1993.

[55] N. Wayne. *An introduction to digital image processing, Prentice Hall.* New York, pp. 129–163, 1986.

[56] A.P. Witkin. Scale-space filtering. *IJCAI*, 1019–1022, 1983.

[57] M. Worboys and M. Duckham. GIS: A computing prespective. *Introductory Readings in Geographic Information Systems*, 2004.

[58] B. Zitova and J. Flusser. Image registration methods: A survey. *Image and Vision Computing*, 21:977–1000, 2003.

Index

T - #0381 - 071024 - C6 - 234/156/14 - PB - 9780367378561 - Gloss Lamination